Project Management for the Creation of Organisational Value

Ofer Zwikael · John Smyrk

Project Management for the Creation of Organisational Value

 Springer

Ofer Zwikael, PhD
School of Management, Marketing and
International Business
Research School of Business
ANU College of Business and Economics
The Australian National University
Canberra
ACT 0200
Australia
e-mail: ofer.zwikael@anu.edu.au

John Smyrk, MEc
School of Management, Marketing and
International Business
Research School of Business
ANU College of Business and Economics
The Australian National University
Canberra
ACT 0200
Australia
e-mail: john.smyrk@anu.edu.au

Additional material to this book can be downloaded from http://extras.springer.com.

ISBN 978-1-4471-5720-5   e-ISBN 978-1-84996-516-3

DOI 10.1007/978-1-84996-516-3

Springer London Dordrecht Heidelberg New York

British Library Cataloguing in Publication Data
A catalogue record for this book is available from the British Library

Library of Congress Control Number: 2010935938

Cover design: eStudio Calamar S.L.

Printed on acid-free paper

Springer is part of Springer Science+Business Media (www.springer.com)

To Maya, Tal and Noa

Ofer Zwikael

To Ruth, Michael, Martin and Carolyn

John Smyrk

Preface

While the profession of project management has matured remarkably, the discipline of project management has tended to lag behind, remaining today essentially as a toolbox (albeit one containing an impressive array of devices and instruments). The irony of this situation is that, although the profession is becoming increasingly preoccupied with the achievement of project management maturity, its own frameworks appear locked into a state of adolescence.

This book seeks to advance an important discussion in which practitioners and researchers are currently engaged related to the foundations of project management in general and its theoretical underpinnings in particular. It does this by proposing a number of new concepts and models that appear to resolve some of the more pressing issues that are of increasing concern to members of the profession.

The ideas and tools presented here should be of interest to three sorts of audience: practitioners who are seeking to understand the strengths and weaknesses of their tools and techniques, academics who need a theoretical scaffold on which to base development of a "theory of projects" and professionals who seek to create a more meaningful and reliable discipline for their work.

The authors wish to acknowledge the contributions towards the ideas presented in this book from academic colleagues, students and business associates. In particular, we mention the input from Dr. Esther Unger-Aviram. We also thank our respective families for their patience, perseverance, encouragement and support over the past few years.

Contents

1 Projects: An Executive Context 1
  1.1   The Role of Projects in Business......................... 1
       1.1.1   Projects: Giving Effect to Strategy 1
       1.1.2   Generating Change with Projects 2
  1.2   The Evolution of a Discipline........................... 3
       1.2.1   Some Historical Themes 4
       1.2.2   Project Management as a Profession 4
       1.2.3   Trends in Today's Project Environment.............. 5
  1.3   Current Issues for Business in Project Planning and
       Management...................................... 7
  1.4   Summary 9

2 The Input-Transform-Outcome (ITO) Model of a Project 11
  2.1   Issues with Current Project Management Methodologies 11
  2.2   Projects as Processes 13
  2.3   Modelling the Project as a Process 15
       2.3.1   The Input-Process-Output Model.................. 15
       2.3.2   Outputs and Outcomes......................... 17
       2.3.3   Target Outcomes.............................. 19
  2.4   The Input-Transform-Outcome Model 22
       2.4.1   Utilising Outputs to Generate Target Outcomes 22
       2.4.2   The Chronology of the ITO Model 25
       2.4.3   Projects and Business Operations 26
       2.4.4   Key Players: Roles, Responsibilities and
            Accountabilities 28
  2.5   Illustrating the ITO Methodology: The Project BuyRite
       Case Study....................................... 30
       2.5.1   The Company................................ 30
       2.5.2   Project BuyRite 32
       2.5.3   The Emerging Shape of Project BuyRite 33
  2.6   Summary 35

3   Achieving Success in Projects   37
   3.1   A Framework for Gauging Performance and Judging Success. . .   37
      3.1.1   Perspectives of Success   37
      3.1.2   The Analysis of Project Performance.   38
      3.1.3   Regression Testing   41
   3.2   Project Management Success   42
      3.2.1   The Conventional Treatment   42
      3.2.2   A New Treatment   46
      3.2.3   Judging Project Management Success   48
      3.2.4   Project Management Success in Practice   50
   3.3   The Worth of a Project   53
      3.3.1   Evaluating a Project's Worth   53
      3.3.2   Measuring a Project's Worth: The Limitations
           of Financial Units   54
      3.3.3   A Project's Benefits   57
      3.3.4   A Project's Disbenefits   58
      3.3.5   A Project's Cost   58
      3.3.6   Calculating a Project's Worth.   59
      3.3.7   Benefit-Cost Analysis and ITO-Based
           Project Assessment   60
   3.4   Project Ownership Success.   62
      3.4.1   Judging Project Ownership Success.   62
      3.4.2   The Regression Test of Project Ownership Success.   63
   3.5   Project Investment Success.   65
      3.5.1   The Regression Test of Project Investment Success.   65
      3.5.2   The Rationale for the Regression Test of
           Project Investment Success   67
      3.5.3   Qualifying Judgements Based on the Regression Test . . .   69
   3.6   Comparing the Three Tests of Success.   69
      3.6.1   Comparing Approaches to Judging Success   71
      3.6.2   Comparing the Many Faces of Success   74
   3.7   Critical Success Processes (CSP).   76
      3.7.1   The Need for an Alternative Critical Approach.   76
      3.7.2   The Critical Success Processes Model   80
   3.8   Summary   83

4   The Project Environment   85
   4.1   Project Phases.   85
      4.1.1   Project Initiation.   86
      4.1.2   Project Planning   87
      4.1.3   Project Execution   88
      4.1.4   Outcome Realisation   88
      4.1.5   Accountabilities During a Project's Life   89

4.2 The Anatomy of a Project 89
 4.2.1 The Elements of a Project 90
 4.2.2 The Types of Work in a Project 93
 4.2.3 A Project's Baseline Documents 94
4.3 Project Governance 95
 4.3.1 Overview of Project Governance 95
 4.3.2 Principles of Project Governance 96
 4.3.3 The Structure of the Project Governance Model 99
 4.3.4 Classes of Entity in the Project Governance Model 101
 4.3.5 Managing the Project Governance Model 108
4.4 The Project Management Office (PMO) 110
4.5 Stakeholder Management 110
 4.5.1 The Concept of Project Stakeholding 112
 4.5.2 The Community of Stakeholders 114
 4.5.3 The Stakeholder Management Process 117
 4.5.4 Stakeholder Identification 118
 4.5.5 Stakeholder Analysis 121
 4.5.6 The Stakeholder Register 123
 4.5.7 Stakeholder Engagement Planning 124
 4.5.8 Stakeholder Engagement Implementation 124
 4.5.9 Stakeholder Engagement Monitoring 125
4.6 The Programme Environment 125
 4.6.1 Related Projects 128
 4.6.2 Coordinated Projects 129
 4.6.3 Staged Projects 131
 4.6.4 Partitioning Projects 132
4.7 The Project Portfolio 132
4.8 Summary 133

5 Starting a New Project 135
5.1 Initiating a Project 135
 5.1.1 The Project Champion 135
 5.1.2 Leading Project Initiation 136
 5.1.3 Conceptualisation 138
 5.1.4 The Role of the Business Case 139
 5.1.5 Developing the Business Case 140
 5.1.6 Expectations, Constraints and Assumptions 143
5.2 Scoping the Project 145
 5.2.1 Setting Project Scope 145
 5.2.2 The Statement of Scope 146
 5.2.3 Identifying and Defining Target Outcomes 148
 5.2.4 Identifying Committed Outputs 152
 5.2.5 Validating Project Scope 152
 5.2.6 Defining Outputs 161

5.3 The Business Case.   162
     5.3.1 The Structure of a Business Case   166
     5.3.2 A Business Case Example   167
     5.3.3 Judging a Business Case   175
     5.3.4 Accepting the Business Case   177
5.4 Appraising Project Risk   177
     5.4.1 The Level of Project Risk   178
     5.4.2 The Effect of Risk on Project Appraisal   179
5.5 Summary   180

6 **Planning a Project: The Roles of the Key Players**   181
   6.1 An Outline of Project Planning   181
        6.1.1 The Need for Planning   181
        6.1.2 The Structure of the Planning Phase   183
        6.1.3 The Outputs from Planning   184
        6.1.4 An Iterative Approach to Planning   188
   6.2 The Project Manager   189
        6.2.1 The Responsibilities of the Project Manager.   189
        6.2.2 Critical Success Processes During
               Project Planning   190
        6.2.3 The Work Breakdown Structure   195
        6.2.4 The Gantt Chart   196
        6.2.5 The Project's Estimated Cost   201
        6.2.6 Risk Mitigation Planning   207
   6.3 The Project Team   214
   6.4 The Project Owner.   216
        6.4.1 Identifying Critical Outputs for
               Close Attention.   216
        6.4.2 Approving the Project Plan   219
   6.5 The Steering Committee.   220
   6.6 Reference Groups and Advisers.   220
   6.7 Project Counsellors   220
        6.7.1 Evaluating the Quality of the Project Plan   221
        6.7.2 The Quality of the Project Plan
               in Practice   222
   6.8 Summary   223

7 **Executing a Project: The Roles of the Key Players**   225
   7.1 An Outline of Project Execution Management   225
        7.1.1 Execution Management Processes.   226
        7.1.2 Accommodating Projects within an
               Organisational Structure.   228
        7.1.3 Top Management Support   232

7.2 The Project Manager 233
    7.2.1 The Project Manager as a Project Execution
        Manager 234
    7.2.2 Communications Management 234
    7.2.3 Risk Control............................... 236
    7.2.4 Issue Management........................... 237
    7.2.5 Schedule Control 239
    7.2.6 The Project Manager's Role as a Team Leader........ 240
    7.2.7 Output Closeout 244
7.3 The Project Team 245
    7.3.1 Formalising Team Roles 246
    7.3.2 Regular Team Meetings....................... 246
7.4 The Project Owner................................ 247
    7.4.1 Managing Scope Change 247
    7.4.2 Scope Change and Risk....................... 249
    7.4.3 Managing Schedule Change 250
7.5 The Steering Committee............................ 251
    7.5.1 A Stylised Reporting Package 251
    7.5.2 A Stylised Agenda 255
    7.5.3 Celebrating Success......................... 256
7.6 Other Key Players................................ 256
7.7 Summary 257

8 Realising Outcomes from a Project: The Roles of the
Key Players...................................... 259
8.1 An Outline of Outcome Realisation..................... 259
    8.1.1 Natural and Synthetic Outcome Realisation 260
    8.1.2 Facilitation............................... 260
    8.1.3 Handover................................ 261
    8.1.4 Outcome Closeout........................... 261
    8.1.5 Evaluation of Project Ownership.................. 263
8.2 The Project Manager 264
    8.2.1 Project Customer Support 264
    8.2.2 Utilisation Monitoring 264
8.3 The Project Owner................................ 265
    8.3.1 Ensuring Effective Utilisation of Outputs............ 265
    8.3.2 Outcome Evaluation 265
    8.3.3 Project Evaluation.......................... 267
8.4 The Steering Committee............................ 269
8.5 Other Key Players................................ 269
8.6 Summary 270

Appendix A: An Integrated Glossary of Project Management
                Terms & Definitions   271

Appendix B: The Input-Transform-Outcome (ITO) Study   325

Appendix C: The Critical Success Processes Study   329

References   351

Index   357

Chapter 1
Projects: An Executive Context

1.1 The Role of Projects in Business

Regardless of the area in which a business operates, its executives and senior managers are under continual pressure to bring about beneficial change. This pressure emerges as an unrelenting demand to undertake projects (as instruments of change) and undertake them well.

Projects and routine business operations are each a type of process, and hence the two share a number of important characteristics, such as resources, work and outputs. However, projects are distinguished from routine business operations by the novelty of the work involved. Whereas projects (such as the Big Dig in Boston) are relatively large and never repeated, business operations (such as the renewal of an insurance policy) are repetitive processes. As a result, projects (and programmes) require different frameworks of management from those that are suited to business operations. In what follows we use the term "business" in its broadest sense, to encompass all formal social endeavour whether it be carried out by the private sector, government, not-for-profit organisations or community groups.

1.1.1 Projects: Giving Effect to Strategy

Although the strategic alignment of projects has become an increasingly important topic of contemporary discussion, not all projects that an organisation funds arise from the demands of a strategic vision. Critical initiatives can emerge spontaneously or opportunistically, including some that may be only indirectly related to current strategy. There are, therefore, three sorts of trigger for projects:

1. *Imposed projects*. Imposed by the environment in which the organisation operates (such as a new law that requires annual safety audits).
2. *Opportunistic projects*. Arising from opportunities to enhance performance (such as acquiring a competitor who is experiencing financial difficulties).

O. Zwikael and J. Smyrk, *Project Management for the Creation of Organisational Value*, DOI: 10.1007/978-1-84996-516-3_1,
© Springer-Verlag London Limited 2011

3. *Strategy implementation projects.* Emerging from a consciously stated strategic imperative (such as a decision to move out of manufacturing into services).

Those projects included in the last group are intended to realise the organisation's strategic plan, while projects in the two other groups have only to be aligned with it. Organisational strategy ensures that the organisation does the right things by pointing it towards its goals. Such goals imply conscious change—all of which must be brought about by projects.

1.1.2 Generating Change with Projects

Regardless of the trigger, each project is undertaken to generate benefits, taking the form of a future flow of desirable (and measurable) end effects. Accordingly, we define a project as a unique process intended to achieve target outcomes. Thus, a project takes on the characteristics of an investment, where resources are purchased today with the prospect of flow of benefits (in the form of target outcomes) tomorrow. This investment interpretation of a project remains useful even if the benefits being sought are non-financial. The entity providing the funds for a project (the funder) is, therefore, more formally described as an investor. Just as returns drive financial investment, benefits drive investment in projects. This conclusion has crucial implications for our later discussion (in Chap. 3) of project success.

While the concept of a benefit is one that most people understand intuitively, weaving it into the tapestry of a project requires skills and techniques that are far from intuitive, as will become obvious from the next two chapters (Box 1.1).

Box 1.1 What is Important to Projects' Funders?

Supporting Research

In a recent study conducted by the authors (see Appendix B), managers were surveyed to analyse the relative importance (to the project funder) of 16 project management factors. Table 1.1 ranks these factors according to their level of importance to project funders.

Table 1.1 The ranking of project management factors as are important to funders

Ranking	Project Management Factors
1.	Achieving target outcomes (benefits)
2.	Approving a business case

(continued)

Table 1.1   (continued)

Ranking	Project Management Factors
3.	Developing a business case
4.	Developing a list of agreed outputs (deliverables)
5.	Producing outputs (deliverables)
6.	Developing a list of agreed target outcomes (benefits)
7.	Effective communications with stakeholders
8.	Monitoring and controlling the project
9.	Developing a project plan
10.	Managing project risks
11.	Assigning a person accountable for target outcomes (benefits) achievement
12.	Support provided by senior managers
13.	Assembling a suitable project team
14.	Updating the project plan
15.	Managing the project team
16.	Developing the project team

This Table shows that achieving target outcomes is the most important factor for project funders. Additional statistical analysis confirms that this high ranking is significant, supporting the claim that funders treat projects as a form of investment

Some care has to be exercised when considering the implications of this ranking. It is clear that various factors have been identified as "important" for quite different reasons. For example, "Achieving target outcomes" would be important when deciding whether to accept a business case, while "Developing the project team" could be important to the execution of an approved project

1.2 The Evolution of a Discipline

An outside observer of the project management discipline would find that over the years, while it has developed many prominent characteristics, none would be as impressive as its sheer formality, reflected for example, in the formation of professional organisations, the adoption of accreditation schemes, the development of recognised educational programmes and the construction of (frequently exquisite) methodologies. Some consider these developments as evidence for a substantial body of knowledge. Others, however, note that while much of this knowledge has many strong, proven and reliable underpinnings, the discipline is at the same time surrounded by a rich and fascinating collage of accepted practice, proprietary products, agreed standards, regularised procedures, anecdotal evidence, folklore, urban myths, professional ritual, assertions, strongly-held beliefs and methodological bias. It is useful to understand the factors that have shaped the project management profession—to understand why it has taken on the form in which we see it today.

1.2.1 Some Historical Themes

We can infer that humans have undertaken "projects" for at least tens of thousands of years. The wonderful cave paintings at Lascaux in France certainly qualify as project outputs under the definitions adopted here. Archaeological evidence that our distant ancestors engaged in both ritual and the creation of decorative artefacts may even suggest that the first projects go back hundreds of thousands of years. It seems doubtful, however, that until the relatively recent past, any recognisable framework of management was adopted for such initiatives. It was not until the complexity of the work increased (for example, when building grand structures such as the great Wall of China) that formality became necessary for the conduct of projects—a formality that was eventually to lay the foundations of an entirely new discipline.

During the 20th century, the complexity of the work itself continued as the dominating problem in project management, and so this became the focus of attention for the profession. For the bulk of the 1900s, project management was preoccupied with solving a range of particularly difficult scheduling and resourcing problems (which, with the support of the Operations Research community, it did very effectively).

In the latter half of the 20th century, the business community began to accept that it too was involved with projects. Business projects are peculiar in that many of their outputs are *represented* by artefacts (rather than being artefacts in their own right). A new business process, for example, is represented by flowcharts and procedures manuals. This class of project also brought with it a new phenomenon—in the form of ambiguous or unclear scope. The lack of a meaningful approach to this problem remains to this day as a key issue for project leadership, planning and management (an issue to which we return a little later).

Today, the interests of the profession have expanded considerably into areas such as risk management, governance, programme and portfolio management and benefits realisation. As the interests of the project management profession continue to widen, there is growing disquiet that the discipline has not dealt effectively with a number of critical emerging issues and, as a consequence, project performance (already the subject of deep concern) will suffer.

1.2.2 Project Management as a Profession

As the range of devices for planning and management has grown, so has the desire for recognition. The establishment of professional organisations such as the International Project Management Association (IPMA) in 1965, the Project Management Institute (PMI) in 1969, and others, has, amongst other things, triggered a desire for professionalism amongst members of the discipline.

With the rise of the professional organisations, project management methodologies, toolkits that contain packaged practices, processes, procedures, instruments and templates have been developed. Although a number of these are extremely popular, (some having given rise to thriving "cottage industries"), there is little in the way of reliable empirical data about their usefulness.

1.2.3 Trends in Today's Project Environment

The nature of projects evolves over time, as does the environment within which they are undertaken. Some of the more noteworthy forces that are shaping the discipline include:

1. *High rates of failure.* Various studies and considerable anecdotal evidence, if taken at face value, all point to a disturbingly high rate of project failure. Despite this, we observe continued high levels of investment in projects. These two phenomena need to be reconciled. There are a number of competing explanations including:

   - Failure rates are indeed high, but the (relatively small) number of winners is more than capable of carrying the (large number of) losers. In other words, the performance of the overall portfolio of projects faced by each funder is adequate.
   - High failure rates are a symptom of a deeper problem that most project portfolios are loss-making. From this, we would be forced to conclude that most decisions in favour of funding projects are wrong.
   - Failure rates are, in fact, relatively low, with most projects yielding an acceptable flow of benefits. In that event, one would have to draw the inference that either there is something wrong with the evidence or the analysis of the evidence is flawed.

   Which of these situations actually prevails? The framework assembled in what follows appears to be consistent with the last explanation. In particular, we show that accepted definitions of project success/failure are incomplete—and hence many studies into failure rates are methodologically flawed.

2. *Disagreement on methodologies and techniques.* Differences amongst some of the existing methodologies are relatively minor. For example, see the bodies of knowledge developed by the PMI, IPMA, Association for Project Management (APM), and to some extent PRINCE2, developed by the Office of Government Commerce (OGC). However, others see the need for more radical action in changing current methodologies. A case in point is 'Rethinking project management' in which a UK based network group is working "… to extend, enrich, reshape and develop this field beyond its current intellectual foundations" (Cicmil et al., 2006). In addition, a number of large and influential organisations (such as Ericsson, Motorola and Philips) have developed their own project

management methodologies. All this has the unintended effect of fragmenting the discipline. Today there is little consensus on the shape of a universal project management methodology. Chapter 2 presents a framework, which is (of necessity) different from those currently in use, because it focuses on effective outcome generation instead on efficient output delivery.

3. *The emergence of customisable project frameworks.* The traditional approach, in which the same processes and tools are applied to all projects, is starting to vanish. The recognition that different project types, industries, cultures, levels of complexity and other factors may influence the way in which the project should be managed, has triggered the development of subordinate project management methodologies. In this context, it is useful to mention extensions to the PMI's body of knowledge for the construction and government sectors and the development of project management approaches for specific cultures (such as the Project Management Association of Japan).

4. *Project complexity.* There is widespread interest in the concept of "complexity" as it relates to projects, but at the same time, there is little, if any, agreement yet about the criteria that make a project complex. There are numerous lines of thought about this, which all appear to be related to the size and/or "definability" of the initiative. While it is clear that there is a lively debate about what makes a project complex, it is much less clear what the implications are for the conduct of projects that have been declared "complex". If it can be shown that the management frameworks for complex and non-complex projects are distinct, then this becomes a significant issue for the profession in general. If, however, it is found that the two sorts of project use the same framework (but to different levels of intensity), then the value added by such a classification is questionable.

5. *Emergence of programmes and portfolios.* There is considerable discussion and debate today about three apparently related but distinct terms: projects, programmes and portfolios. The concept of a portfolio, already well-established in other disciplines such as finance, corporate strategy and economics, has also been introduced to the project management discipline. As suggested in the next section, the accepted meanings of this term are readily adapted to accommodate the concept of a portfolio of projects. Programmes are more problematic. Unfortunately, that particular debate has been obscured by semantic confusion. As defined here, a project is a simple concept with a structure that includes resources, work, deliverables and benefits. As simple as the concept of a project might be, it applies, with uniform relevance, to projects of any scale. Some of the definitions of programmes that have been proposed by members of the profession quickly collapse into a rather unhelpful argument about word preferences. Our coverage of this topic takes a view that the programme/project argument is little more than a relatively straight-forward discussion about how related projects should be coordinated. The relationships amongst projects, programmes and portfolios are discussed in Chap. 4.

6. *Globalisation.* Because of the globalisation of business, many organisations face projects that are spread over different geographical locations and different

time zones—involving team members with a variety of nationalities. International projects, once a rare phenomenon, are now commonplace. When projects involve multiple stakeholders from different countries, the task of managing them becomes even more complex, requiring a deep understanding of cultural diversity. The phenomenon of virtual teams is, in many respects, a response to the issues created by global projects.

1.3 Current Issues for Business in Project Planning and Management

Accepted project practice and the environments within which projects are undertaken present a range of key issues for executive management and members of the profession alike. As summarised below, these issues are interdependent—each one is intimately tied up with the others. Our approach therefore is to assemble a general framework before attempting to address them.

1. *Myopia in the traditional view of a "project".* Up until the end of the 20th century, a project was generally seen as a process to produce agreed outputs fit-for-purpose, on time and within budget. Since then we have seen a growing realisation that this view is incomplete—with increasing numbers of researchers and practitioners accepting that projects are intended to realise benefits. The implications of such a principle are profound. Three in particular are noteworthy:

   - Judgements about success/failure should account for benefits, and yet, at the same time, it would appear cavalier in the extreme to abandon the conventional criteria of scope/time/cost.
   - If judgements about success require measurement of benefits and if benefits are realised after outputs are delivered, then a project must continue beyond delivery of its outputs until the point when its benefits are realised, or at least secured.
   - Someone in the funding organisation should be made accountable for the generation of benefits from a project. Despite wide acceptance of this principle, the "benefits-realisation" models currently being proposed, do not explain satisfactorily how projects, outputs and benefits are related, nor do they appear to offer techniques that can systematically and consistently rank projects as candidates for investment.

   We propose a conceptual model that not only separates outputs and benefits in a rigorous way, but that also explains the mechanism by which benefits are generated by a project.

2. *Unreliable (or non-existent) statements of project scope.* Even if we ignore the role of benefits in project definition (and accept the conventional outputs-based view), it is obvious that before work gets underway two questions must be answered: "What outputs are to be delivered?" and "What characteristics are to

be built into each output?". Expressed another way: "Of all the lists of outputs (and required characteristics) that might be proposed, which one defines the scope of this project"? This is what we call the scoping problem. The existing literature is almost silent on the issue, either ignoring it completely or offering little guidance on how the appropriate decision is to be taken. This is surprising because the selection of outputs will not only determine the project's costs and timeframe, but it will also have a fundamental impact on later generation of benefits. We propose a methodology and tool to support the scoping process and solve the scoping problem.

3. *Commitment to infeasible projects.* A project is infeasible when its outputs cannot be produced, delivered and implemented within the imposed constraints of time and cost. The evidence (both empirical and anecdotal) suggests that impressively large numbers of projects are undertaken that are later shown to be infeasible. Clearly, the approval of a project that is known to be infeasible would represent perverse behaviour at best and professional negligence at worst. This then begs the question, "Why are infeasible projects accepted in the first place?" We suggest that infeasible projects are undertaken in organisations because of other forms of delinquency. Firstly, timeframes are seen (mistakenly) as being only vaguely related to the work of a project—and so it is believed they can be set arbitrarily. Secondly, there is a naive belief (often amongst senior stakeholders) that the achievability of a timeframe is related to its urgency. "The deadline is too important to miss!" as if the more disastrous the consequences of missing the date, the less likely it is to be missed. It may well be too important to miss, but what if, at the same time, it simply cannot be achieved? There is also confusion between the processes of estimation and negotiation. "The budget is non-negotiable". It might well be non-negotiable, but what if, at the same time, it is inadequate? To deal with these issues we propose that project timeframes and costs must be accompanied by "evidence of achievability" before they are accepted. It is the responsibility of senior stakeholders to ensure that projects are feasible before they are approved. Chapter 6 discusses the processes that should be undertaken to confirm a project is feasible for execution before any commitment is undertaken.

4. *Breadth of stakeholding.* Every project has stakeholders who can influence (or be influenced by) its results. In some cases, these players may have different expectations about the project. For example, project customers (a defined term in the glossary provided in Appendix A) may expect the project to address concerns that they see as important, but in which the funder has no interest. In other cases, there may be conflicting and irreconcilable views amongst stakeholders about the project, including some who may even oppose it. This could very well be the case, for example, with employees whose roles will be changed or who might even face the loss of their jobs. The greater the number of stakeholders in the project (and the greater the spread of their interests in the project), the greater the effort required to engage them successfully. Because project success may be significantly impacted by the behaviours, attitudes and involvement of stakeholders, frameworks of project management usually

include components that seek to influence the form, direction and nature of that impact. Chapter 4 explores the concepts and techniques that underpin stakeholder management.

5. *Inadequate governance models.* Much accepted practice is based on simplistic, informal approaches that experience tells us, in general, simply do not work. A case in point is the way projects are organised. Poorly thought-through organisational arrangements actually work against projects and can contribute to poor performance or failure. Project governance is concerned with the deployment of key players in workable organisational models that are free of the worst problems created by ad hoc approaches. Project governance modelling (together with its supporting principles) are all discussed in Chap. 4.

6. *The need for clarity about project accountabilities.* The central role played by outcomes in the assessment of projects is a key thrust of the framework presented here. A proposed principle of management relates to the assignment of accountability for each criterion that will be used eventually to make judgements about a project. The models presented here confirm that, while it is appropriate to make the project manager accountable for many of these (in particular for the delivery of outputs that are fit-for-purpose, on time and within budget), it is not appropriate to make him/her accountable for securing project target outcomes. Accountability for target outcomes should be assigned to the project owner (a role not recognised, let alone defined, in the conventional view of a project). Project accountabilities are discussed in Chap. 4.

7. *The project manager's place in the organisational structure.* The traditional "functional organisational structure" often places project managers in the position where they lack the power, resources, budget and authority necessary to influence a project's results. This model does not serve projects well. Because, for many organisations, project-oriented activity is still a relatively low proportion of overall business load, a project-based structure is not appropriate. As a result, many project managers face some difficult challenges, such as potential power clashes with line managers, contention for the time of team members who face dual lines of reporting and conflicts between authority and responsibility. This issue, too, can be addressed through project governance models (discussed in Chap. 4).

1.4 Summary

The project environment is shaped by many factors. The project management profession has come a long way, especially in the creation of methodologies, tools and techniques to address the challenges faced in this environment. However, some major shortcomings with practice and theory are becoming evident as our interest in projects moves beyond outputs towards outcomes. These will be discussed in Chap. 2, where an alternative framework for projects is proposed.

Chapter 2
The Input-Transform-Outcome (ITO)
Model of a Project

2.1 Issues with Current Project Management Methodologies

The need for a new project framework emerges from various issues and criticisms related to current project management methodologies (especially those raised by the research community), as outlined in the following discussion:

1. *The neglect of project benefits.* Existing project management methodologies are primarily concerned with output delivery, often neglecting outcome and benefit achievement (Dvir, 2005; Fraser, 2003). As a result, many organisations have become preoccupied with *efficient* output delivery (for example, by completing projects on time) instead of *effective* outcome generation (by seeking to generate desired outcomes). Benefits (defined as the flows of value that arise when desirable outcomes are achieved from a project), are in turn neglected (because they too are related to project effectiveness rather than project efficiency). The methodology proposed here not only highlights the central role of project benefits but also identifies processes that contribute directly to their successful achievement.

2. *Scenario-specific frameworks.* An analysis of the approaches adopted by various professional organisations reveals major differences in processes and terminology across industries and cultures. For example, the information technology sector has developed particular project management methodologies and an associated lexicon. Moreover, formal project management roles, such as "planners" and "estimators", are widely accepted in the construction sector, but not recognised in other industries. Cultural diversity has also triggered the development of unique project management methodologies in various societies. This suggests that current project management methodologies have to be adapted to different project scenarios because they are not generic and lack robustness (see also Dvir, Sader, & Pines, 2006). We propose in this book a rigorous project management framework that can be applied in all project contexts, while remaining flexible enough to accommodate the peculiarities of each project.

O. Zwikael and J. Smyrk, *Project Management for the Creation of Organisational Value*, DOI: 10.1007/978-1-84996-516-3_2,
© Springer-Verlag London Limited 2011

3. *Inconsistent and incomplete terminology.* The terms used throughout the project management profession have not been standardised. As well as different words being used to identify the one concept, particular terms are also used to identify unrelated concepts. For example, the "project customers" in the service sector, are usually called "end users" in the information technology sector, while "sponsor" can refer to any of a number of stakeholders such as, funder, owner and champion. There is also considerable confusion with use of "outcome" and "outputs" in various sources (Nogeste & Derek, 2005). All the terms used throughout this text have the meanings offered in the integrated glossary provided as Appendix B (Box 2.1).

As part of the attempt to develop a rigorous (but flexible) theoretical framework, we propose conventions and rules for the consistent labelling of the terms and concepts we use. One of these conventions involves *word structure*, particular rules governing the way that terms are labelled. For example, outputs are labelled with nouns ("a new business process", "a new bridge") while outcomes are usually labelled with participial adjectives ("increased sales", "reduced incidence of domestic violence").

The following sections describe an approach which we claim not only addresses these issues, but a number of others as well.

Box 2.1 Concerns about Project Management Theory

From the Literature

Many scholars criticise current project management methodologies and the lack of a robust theory. For example, Shenhar and Dvir (1996) argue that "most research on the management of projects is relatively young and still suffers from a scanty theoretical basis and a lack of concepts". Meredith (2002) claims that the project management literature is often characterised by non-rigorous research methods and frameworks that are unrelated to previous work. Various reasons for the relative immaturity of project management theory have been proposed. Specific criticisms include claims that the literature has been practitioner-driven (Jugdev, 2004) and reliant on "war stories" (Meredith, 2002), extensive use of normative (rather than positive) approaches, and appeal to lists of factors derived from surveys of project practitioner opinions, rather than empirical research grounded in theory (Packendorff, 1995). Bygstad and Lanestedt (2009) support this claim view by noting that project management methodologies are rarely used "as is" in Germany and Switzerland, but usually modified or adapted before application. Increasingly, there have been calls for improved theory generation through research designs that build on existing literature to develop models for rigorous, evidence-based testing in industry (Meredith, 2002).

2.2 Projects as Processes

All work (both project and operational) can be viewed as a collection of processes. A process is a structure of activities that produce an identifiable output. An output always takes the form of an artefact. Examples come readily to mind: a new insurance policy, the Sydney Opera House, a fishing licence, the prototype of a new chemical pump, a reengineered procurement process or an order (placed on a supplier). Because outputs, also known generically as *deliverables,* are artefacts, they are always labelled with nouns. In some cases, outputs take the form of a change to an existing artefact. Consider a process that will see the personal details of an insurance policy holder updated. That output would be appropriately entitled "Updated policy".

A closer look at the illustrative list of outputs given above suggests that they emerge from processes of two kinds: those that will be done only once and those that will be repeated. Projects are unique processes, while business operations are repeated. Processes of both kinds share many characteristics: they consume resources, involve work that can be systematically described by some sort of script and they produce outputs. Despite this similarity, distinct frameworks of management have evolved for projects and business operations.

Frameworks for managing projects (rather than operational processes) are the focus of this book. It will become clear that, because these frameworks involve significant work and resources of their own, their suitability for the management of "day-to-day" processes is problematic. Instead, "business-as-usual" processes yield to different approaches, collectively described as *operational management.*

Operational processes and projects differ in many respects, one of the most important of which concerns "change". Not only are projects intended to introduce specific, defined, targeted change into the world, but the extent to which this actually happens determines whether a project was successful. By way of contrast, operational processes are, in a sense, intended to maintain the status quo. Because different frameworks of management have evolved in business for the two classes of process, the question arises "under what circumstances is it appropriate to execute a process under an operational management framework, and under what circumstances is a project management framework required?". The following test is proposed. If the work being undertaken already has a reliable, comprehensive script, then we would execute it according to that existing script—in other words, as a business process. Reliable scripts can only be obtained by refining them over many repetitions of the same piece of work. For example, in many jurisdictions, the registration of a new car is described in a detailed script that has been developed progressively by state road authorities over long periods of time. On the other hand, an exercise to re-engineer the registration process (and, accordingly, introduce major changes into the existing script) is a unique exercise and so should be managed as a project.

If a script does not exist, then the piece of work can be executed in only two ways: by writing a (somewhat tentative) set of instructions, which is then followed during execution or by "making it up as you go along". Writing a script for a novel piece of work is, in fact, the foundation of *project planning*. Because we do not really expect that a new script (especially if it is assembled *ex nihilo*) will be correct in every detail, it is necessary to monitor its execution (to detect any errors) and then correct it whenever it is found wanting. Monitoring and correcting a tentative script for a process is the foundation of *project management*.

Now what about "making it up as you go along"? This implies execution without a script. In general, the more important the work the less attractive this approach, however anecdotal evidence suggests that very small pieces of work (such as taking a phone call) are best handled in this way. Such processes are identified here as ad hoc *tasks*.

From this perspective then, processes can be categorised into three, according to the particular set of "management rules" we use to guide their execution:

1. *Business (or operational) process*. If a piece of work has a reliable script, run it as a business process (using the practices of operational management). Operational management practices are generally applied to repeated, "relatively complex" processes.
2. *Ad hoc task*. If a piece of work has no script and it is "small", execute it as an ad hoc task (by making it up as you go along). Treatment as an ad hoc task is suited to all "relatively simple" processes, regardless of their novelty.
3. *Project*. If a piece of work is "large" and has no script, run it as a project (using the practices of project management). Project management practices are suited to novel processes and become more effective as the complexity of the work increases.

To apply these rules, it is not necessary to define "complex", or "small" because their role is not to classify work, but to simply guide the selection of an appropriate management approach.

Figure 2.1 suggests a mapping from process novelty and complexity to an appropriate choice of management regime. It should be noted that neither axis in this diagram has a defined measure and so it is purely indicative.

Fig. 2.1 Classifying the management of work according to its novelty and complexity

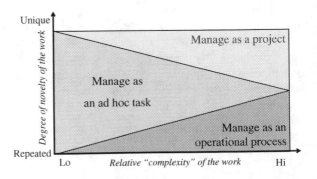

2.3 Modelling the Project as a Process

This section describes the "conventional" Input-Process-Output (IPO) model, (which underpins many project management methodologies), discusses a shortcoming of this approach and presents the foundations for the extended project methodology that follows. In the course of this discussion it becomes necessary to consider a number of entities who have various "interests" in the project, otherwise known as stakeholders. A more comprehensive discussion of stakeholders is provided in Sect. 4.5 in Chap. 4.

2.3.1 The Input-Process-Output Model

We have noted that projects are a subset of processes. As the work of a project is executed, it consumes economic resources, generically identified as *inputs*. For convenience in this discussion, resources are divided into two categories: the labour (of the project team) and the money (outlaid to obtain all purchased-in resources).

When linked in a diagram, inputs, processes and outputs provide us with a simple but extremely potent conceptual view of a project. The relationship among these terms is known as the Input-Process-Output (IPO) model. This model (used extensively in the operations management arena) is shown in Fig. 2.2.

All processes (including ad hoc tasks, business operations and projects) have an underlying IPO model. The IPO model implies a chronology (left-to-right) where, in turn: resources are made available for the work of the project, the work is executed to produce certain outputs and those outputs are then delivered to the outside world. The IPO model itself also figures prominently in project management methodologies themselves, such as the Project Management Institute's Project Management Body of Knowledge (PMBOK) where each of 42 recognised project management processes is described in terms of: the required "inputs" for its effective execution, the expected "outputs" from this process and the "tools and techniques" to guide the process (Box 2.2).

Fig. 2.2 The input-process-output (IPO) model of a project

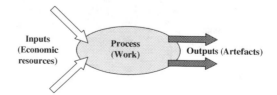

Box 2.2 "Everything is a Project"

A Popular Confusion

It is common to hear the claim that "Everything is a project", with the implication that all of the frameworks and protocols surrounding a project should be applied to all work. Sometimes this is taken even further, with organisations embarking on rather problematic initiatives to "projectise" all their operational processes.

A little thought reveals that, although usually well-intentioned, exercises of this kind do not make a great deal of sense. Take an initiative to outsource an organisation's IT services. This (obviously large) exercise demands, amongst other things, that a business case be tabled so that the work (and expenditure) involved can be approved. (As will be seen in Chap. 4, a business case contains all of the essential information so that a reliable decision can be made about funding the exercise). *Tabling a business case* is therefore an essential element of approval for doing the work of a project. Consider, by way of contrast, the work involved in processing a customer's trolley at the checkout of a supermarket. It would be a cause of some surprise if, on arrival at the checkout, the staff member announced that it would be necessary to have a business case approved before serving you! Clearly, unlike projects, repeated operational processes do not require approval whenever they are executed. (This general proposition is in no way weakened by the fact that, at some time in the past, there may well have been a project to design the current checkout process and that this project had a business case).

And yet, both forms of work are (at least in certain critical respects) similar—they consume resources, demand work and produce outputs. Therefore, both have an IPO model. Presumably, we could extend this list of shared features even further to include many other components such as a framework of risk management. So what can we conclude from this similarity? All we can infer is that regular day-to-day business operations and projects are both examples of *processes*. In other words, *all work is a process*. Clearly although "business-as-usual" is a special case of a process, and a project is a special case of a process—the conclusion (that everything is a project) is false.

Now having said all that, it does indeed make a lot of sense for organisations to formalise the frameworks of management they use for their routine processes, but in doing so they should be more concerned with "operationalising" these processes rather than "projectising" them.

2.3.2 Outputs and Outcomes

Outputs are produced by the work carried out during the execution of a process and are of two kinds:

1. *New artefacts*, where none existed before (such as a bridge over a river).
2. *A change to an existing artefact*, such as a refurbished office building. For ease of identification, we label this particular class of output as an "alterant" (although the term has no particular analytical significance).

Clearly, this is not a dichotomy because, given enough changes, any existing artefact can eventually be treated as new. The word "new" is understood (even if it is not used explicitly) as suggested by examples such as: house, bridge and report. Examples of alterants are: repainted house, repaired bridge and amended report.

Outputs can be viewed as an important *result* to emerge from a project, but an even more important class of result takes the form of the end-effect generated as a consequence of the implementation of those outputs. Such results are called *outcomes*. Although outcomes are not displayed in the IPO model, they represent the very reason for producing outputs, and hence bear further discussion. Outcomes can be grouped in various ways. One important classification indicates whether the outcome is desirable or undesirable. Desirable outcomes can be further categorised as targeted or fortuitous. *Target outcomes* are consciously sought at the outset and represent the rationale for funding the project. *Fortuitous outcomes* are desirable end effects that emerge from a project despite not being targeted.

Target outcomes can be isolated with three simple questions:

1. What is the project's purpose?
2. Why is each of the project's outputs being produced?
3. What end result is the funder (the person approving the allocation of resources to the project) expecting from the exercise?

Consider four illustrative projects: the construction of a new office block, the drawing up of a supply contract with a major customer, the re-engineering of an existing business process and the development of a prototype chemical pump. In each case a rationale for the work involved can be uncovered by asking the questions posed above. An office block may be built to generate a flow of rental revenue. A contract may be drawn up to reduce the risk of supply interruption or price uncertainty. A business process may be reengineered to lower operating costs. A prototype pump may be developed (and evaluated in service) so that a decision can be taken on releasing a new product to the marketplace.

We now draw a subtle, but critical, distinction between the concepts of "tangibility" and "measurability". While an output is always a tangible artefact, an outcome is always a measurable effect. Tangible means, "can be touched", and so

Table 2.1 Outputs versus outcomes

Characteristic	Output	Outcome
Intent	What is to be delivered?	What is the objective?
Form	Artefact	Measurable end effect
Specification	Establish critical features and characteristics	Set seven attributes (characteristics)
Labelling	Noun	Participial adjective
Creation mechanism	Production or delivery	Generation or realisation
Certainty	Production can be guaranteed	Generation cannot be guaranteed
Manageability	Production can be controlled	Generation can only be influenced
Measurement	Through critical features and characteristics measured in quality tests	Through one or more agreed measures with agreed units and dimensions
Tangibility	Outputs are tangible	Outcomes are intangible (but measurable)
Appearance	Impossible without execution of process	In certain cases possible, even if process is not executed
Lead time	Available immediately after process is executed	In general, delayed until after execution of the process

tangibility is a required attribute of an output. Measurable means "can be measured" and is a required attribute of a target outcome. However, because they are end effects (rather than artefacts) outcomes are not tangible. Outcomes can always be expressed as a change in the value of a variable associated with an end-effect, for example: "reduced waiting times", "increased market share" and "compliance with new legislation". Table 2.1 summarises the differences between outputs and outcomes.

For reasons discussed further in Sect. 2.4, outputs have a high degree of certainty (whereby it is reasonable to assume that, if technically feasible, they will, in due course, appear), while outcomes are characterised by uncertainty (because, in particular situations they might not appear). These differences are reflected in the terminology we use where outputs are (variously) *produced*, *delivered* and *implemented*, while outcomes are achieved, generated or *realised*.

In general, different stakeholders view the same outcomes from a project as having different values. "Value" in this sense equates to a level of desirability. If an outcome is judged as having a positive value, it is called a benefit; if it is negative, it is called a disbenefit. (It is possible, of course, that one outcome from a project could be viewed as a benefit by one stakeholder and as a disbenefit by another). As is discussed later, outcomes and benefits are distinct but intimately related, because outcome generation drives benefit generation.

2.3.3 Target Outcomes

To understand why target outcomes can usually be entitled using participial adjectives ("increased ...", "decreased ..." and so on), a more detailed discussion is required. Consider a project that is being promoted to reduce traffic volumes in a city's Central Business District. As proposed, the project's outputs include a cross-city tunnel, changes in the configuration of existing city streets, a tolling system, a suite of management/maintenance processes and a new business unit to operate the facility. Three scenarios surround this project, each one describing how the world might be shaped (or at least that part of the world relevant to the project):

1. A *"Now"* scenario. Describing the current position in which we now find ourselves. This relates to the present state of affairs, characterised by the actual values taken by certain measurable variables-of-interest such as congestion, noise, air pollution and pedestrian accident rates.
2. A *"Yes"* scenario. Describing a future position in which we would like to find ourselves if the funder approves the project. This relates to the desired state of affairs, where the same four variables would take on targeted (and presumably different) values in response to the project being approved and completed.
3. A *"No"* scenario. Describing the future position in which we would find our-selves if the funder rejects the project. Like the other two, the "No" scenario would also be characterised by peculiar values for each of the four variables. These values could well be different to those found in either of the other two scenarios.

In summary then, there are potentially three different values ("Now", "No" and "Yes") for each of the four environmental measures (congestion, noise, air pollution and pedestrian accident rates). Figure 2.3 illustrates the relationships amongst the three scenarios for the vehicular tunnel project.

Target outcomes are the differences in those variables selected to characterise the "No" and "Yes" scenarios. It is important to note that the target for a desired outcome is found as the difference between "No" and "Yes", it is not found as the difference between "Now" and "Yes".

In light of this definition, outcomes are not time-related effects. In other words, "reduced pollution" does not mean (necessarily) that pollution levels will fall because of the tunnel, it means that they will be less than the levels experienced had the tunnel not been built. Counter-intuitively, it is quite conceivable that the project is outstandingly successful even though pollution levels have actually risen. Despite this, in practice, project participants will often express desired outcomes as the difference between the "Now" and "Yes" scenarios (Thus creating the illusion of a time-based target outcome). When this happens they usually assume that without intervention, the "Now" scenario will remain, eventually unfolding as the "No" scenario. There will be circumstances where that is a reasonable expectation, but there will also be situations where the "Now" scenario is better used as a surrogate for the "Yes" scenario. Consider a car maker who

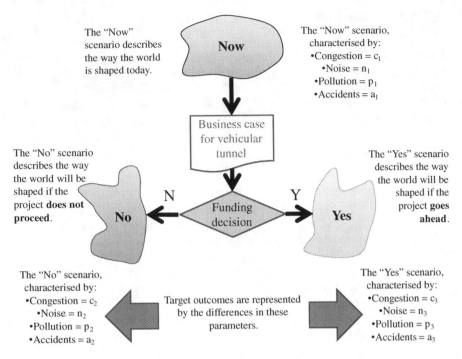

The "Now" scenario describes the way the world is shaped today.

Now

The "Now" scenario, characterised by:
- Congestion = c_1
- Noise = n_1
- Pollution = p_1
- Accidents = a_1

Business case for vehicular tunnel

The "No" scenario describes the way the world will be shaped if the project **does not proceed**.

No

N

Funding decision

Y

Yes

The "Yes" scenario describes the way the world will be shaped if the project **goes ahead**.

The "No" scenario, characterised by:
- Congestion = c_2
- Noise = n_2
- Pollution = p_2
- Accidents = a_2

Target outcomes are represented by the differences in these parameters.

The "Yes" scenario, characterised by:
- Congestion = c_3
- Noise = n_3
- Pollution = p_3
- Accidents = a_3

Fig. 2.3 The Now/Yes/No scenarios for the vehicular tunnel project

currently enjoys tariff protection in the domestic market, but who is faced with the removal of trade barriers as part of a free trade agreement. The firm's existing market share of 50% is expected to fall to 45% as a result. A marketing initiative is undertaken with the objective of retaining market share. Here, "Now" and "Yes" are the same (in terms of market share), while "No" is 45%. Thus, the target outcome of the venture is, quite properly, expressed as "increased market share" (of 5%) (Box 2.3).

Because target outcomes are associated with *desirable* end effects, it is tempting to express them as an "improvement" in some variable. For example, in the case of the Cross City Tunnel, "decreased congestion" might be described as "improved traffic flow". It is undesirable to use "improved" in the title of a target outcome for two reasons:

1. It is unnecessarily vague: if a measure of a variable has improved, it has changed. If it has changed, it has either increased or decreased. We can conclude therefore that "improved" can always be replaced with "increased" or "decreased".
2. It is used in the title of certain outputs: (defined above as "alterants"). For example, a key output from a project to reduce an organisation's purchasing costs might be "an improved procurement process".

Box 2.3 Scenarios, Outcomes and Operational Processes

Illustration: Hospital Process Re-Engineering

Consider a project to improve the public health service through the reengineering of core hospital-related business processes. Two desirable changes would be "reduced waiting time (for certain benchmark procedures)" and "reduced average costs" (for those same procedures). An undesirable change might be "increased staff turnover". In this example, the "Now" scenario is driven by current practices and procedures that cause the waiting time and cost variables to take on a (presumably) high value. The "Yes" scenario is envisaged as a world with different practices and procedures that cause the waiting time and cost variables to take on a (desired) low value. The "No" scenario could (in this example) represent a world in which current practices and procedures are being overwhelmed by the load of an aging community and so the waiting time and cost variables might be expected to grow even larger. In general, then, the variables that are targeted for change by a project will be the same variables that characterise the way certain operational processes behave. In the language of the business process engineering profession, the variables that define the target outcomes from process improvement initiatives always take the form of changes in specific process metrics.

A critical measurement issue now emerges, whereby if a project is undertaken, the "Yes" scenario will be revealed and the "No" scenario will remain unknown. If the project is not undertaken, the reverse is true. How then are outcomes to be targeted? Setting values for target outcomes has to be done before a decision is taken on the project and so the situation appears even more tenuous because, at that point, neither of the eventual "No" and "Yes" scenarios has been revealed. Outcomes are set by *predicting* values for the variables that characterise the "No" and "Yes" scenarios (and then taking the difference between these predicted values for each target variable). To deal with this issue, it is necessary to use some form of projection in a business case so that the values of the variables used to define target outcomes can be produced for both the "No" and "Yes" scenarios. In most cases the techniques required will be very obvious (such as simple "What-if" analysis), but in others they may involve very sophisticated tools such as simulation analysis and statistical modelling.

A fundamental question in the discipline of project management is "Why do we invest in projects?" If the answer is restricted to an IPO view of the world, it appears to be "To deliver outputs", but that simply begs the question "Yes, but why do we want those outputs?" The "real" answer clearly has to do with the end effects that those outputs can bring about. We invest in projects to generate desired outcomes because desired outcomes will, in turn, generate a "flow of value" (a benefit) for particular stakeholders (called the project's "beneficiaries"). It should be noted that a funding organisation does not itself have to be a beneficiary to make a rational decision in favour of a project, as would be the case for example, of a local government that funds a streetscape beautification project for the benefit of nearby restaurants and their patrons. Loosely, outputs are the "means" to an "end" (in the form of outcomes and, eventually, benefits). Because it does not recognise outcomes, the IPO model is clearly inadequate for effective project management and so an alternative (based on an extension to the IPO view) is presented in the next section.

2.4 The Input-Transform-Outcome Model

In this section, we present the ITO model, which serves as our foundation theory.

2.4.1 Utilising Outputs to Generate Target Outcomes

The above discussion reveals a serious shortcoming in the IPO model (at least as far as a representation of a project goes) in that it makes no mention of outcomes. To address this problem, we modify it by adding two elements (utilisation and outcomes) to the three that are already there (inputs, process and outputs). The resulting five-element structure represents the ITO model of a process, as shown in Fig. 2.4. The ITO model is so-named because it seeks to explain how Inputs on the left are Transformed into Outcomes on the right (Smyrk, 1995).

The left hand half of the ITO model is simply the IPO model, to which has been appended a utilisation mechanism and a *flow* of outcomes. The original "left-to-right" chronology can now be extended. The project's outputs are eventually delivered to someone who then utilises them in a way that subsequently contributes to target outcomes. The entities who utilise a project's outputs in this fashion are called the *project's customers* (not to be confused with the *organisation's* customers, the project's *beneficiaries* or the project's *funder*). While every

Fig. 2.4 The ITO model of a project

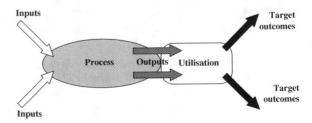

execution of a process has an IPO model, not all executions of all processes have target outcomes. In some cases, there is no meaningful ITO model for individual executions of the process in question. This is a relatively common situation with operational processes. All projects on the other hand, have target outcomes (because they are designed to effect some intended change) and so for most projects, (even those where target outcomes have not been stated explicitly) it is possible to infer an ITO model.

In the left-hand (IPO) part of the ITO model, there is a two-way implication between the process and its outputs. This means that the outputs will exist if, and only if, the process is executed. Because we treat processes as controllable, we are, therefore willing to guarantee outputs. Consider a project to replace an aging narrow steel truss bridge with a wider new concrete girder design. The bridge runs across a river that separates a heavy industrial estate from a harbour that handles both general cargo and containers. A contractor for such an exercise would normally be quite comfortable guaranteeing the replacement because the work involved can be controlled to a high degree.

By way of contrast, in the right-hand part of the figure, the link between utilisation and target outcomes is much weaker. The generation of outcomes is merely correlated with utilisation (and, by implication, with the production of the project's outputs). In other words, it is conceivable that for a particular project, either of two scenarios could unfold: despite the utilisation of outputs, target outcomes are not generated or (more startlingly) target outcomes might well be achieved, even if the project does not proceed.

In the case of the bridge replacement, assume that the target outcome is reduced travel time, with a threshold of 25%. There are many factors that could cause this result not to be achieved (such as an unexpected increase in the number of wide-load trucks that were previously banned from using the steel bridge). There are also many factors that could cause the desired reduction in travel time to occur even if the project is not undertaken (such as the closure of the general cargo terminal). The possibility of both these scenarios confirms that while construction of the bridge can influence the likehood that target outcomes will be generated, the bridge replacement is neither necessary nor sufficient to achive that result. Together, the list of target outcomes, together with the list of its outputs, defines the project's *scope*.

A further observation about utilisation is appropriate. There is a peculiar (and relatively uncommon) class of projects where the realisation of target outcomes is completely independent of any utilisation of outputs by a customer. An example of such a "non-utilisation" outcome is Ripple Rock (see Box 2.4). We distinguish outcomes that arise without the need for utilisation from those that require utilisation of an output by identifying them as "natural" and "synthetic" respectively. While it is tempting to claim that natural outcomes are generated "automatically", in general, this is not true. It is conceivable that, in certain projects, a desired natural outcome might not be generated to a desired level (indeed, in extreme cases it might not happen at all). All we can say about natural outcomes is that they occur without anyone utilising the project's outputs and they can be generated below, at, or above the targets set in a business case.

A number of conceptual devices that seek to explain the relationships amongst inputs, outputs and outcomes have been proposed, including the Logic Model (Kellogg Foundation, 2004) and the Outcome Profile™ (Walker and Nogeste, 2008). While these models accept that inputs, processes, outputs and outcomes appear in that order, they offer no mechanism to explain what "causes" outcomes to appear. Similarly, PRINCE2 (OGC, 2007), a popular proprietary project management methodology, also highlights the importance of outcomes, but again, appears to ignore both outcome generation and tools to facilitate the

Box 2.4 The Peculiar Case of Projects with Target Outcomes but No Utilisation

Illustration: Ripple Rock

Ripple Rock was a dangerous undersea pinnacle (which rose to within 3 m of the water's surface in the Seymour Narrows near Campbell River, British Columbia in Canada (Wright, Carpenter, Hunt, & Downhill, 1958). Enormous tides sweeping over the reef created extraordinarily chaotic and dangerous sailing conditions that, over the years, are reported to have sunk more than 120 vessels with the loss of 114 lives. After a number of unsuccessful attempts to deal with the problem, in 1958, a project was undertaken to remove the reef. This was achieved by drilling down through nearby Maude Island, horizontally under the bay and then vertically up into Ripple Rock itself. There a network of "coyote" tunnels was filled with over 1,270 tonnes of high explosive. Ripple Rock was then destroyed in, what was at the time, the largest ever non-nuclear peacetime blast.

An example of the many valid outcomes and outputs that could be used to define this project are:

Target outcome: reduced risk exposure (to loss of vessels and lives in the Seymour Narrows).

Output: removed/destroyed reef (in the terminology introduced earlier, an "alterant").

In this case, no one actually utilises the "removed" rock. Any reduced risk exposure that is attributable to the reef will be the result of its destruction. Accordingly, using the terminology introduced into the ITO methodology we note that, although this project has beneficiaries, it has no customers who utilise outputs. Accordingly, "reduced risk exposure (to loss of vessels and lives)" is an example of a "natural" outcome.

Fig. 2.5 suggests a diagrammatic representation of an ITO model where there is no utilization, as is the case with the Ripple Rock project.

Fig. 2.5 The ITO model of a
project with no utilisation

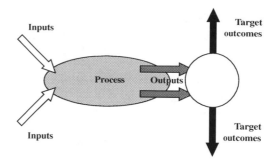

transformation of outputs to outcomes. Moreover, because it terminates with a
"closing a project" phase, there is no description of the processes that should be
executed during outcome realisation.

2.4.2 The Chronology of the ITO Model

The chronology that underpins the ITO model can be made a little more explicit by
showing a horizontal timeline. It does not really qualify as a true "time" axis
because the elements in the ITO diagram are not drawn to a time scale. Under this
view, the "work" part of the model (represented by the process ellipse) would
have defined start and finish dates, notionally obtained by dropping perpendiculars
from the left and right hand extremities of the ellipse onto the X-axis, as shown in
Fig. 2.6. T_1 represents the date on which the production of the project's outputs
begins, while T_2 indicates the date on which the last output is implemented.
The difference between these two dates is the duration of the work required to
produce, deliver and implement the project's outputs. This corresponds to the
conventional concept of project duration (e.g. PMI, 2008). Under the ITO model,
however, the duration of the overall project is longer, continuing after utilisation
has begun until the flow of outcomes has been secured, as indicated by T_3.

Figure 2.6 shows the most general situation, whereby utilisation takes place
over an indefinite period into the distant future. In the example of the Cross City
Tunnel (Sect. 2.3), the outcome "Reduced congestion on surface streets" may be

Fig. 2.6 The chronology of
the ITO model

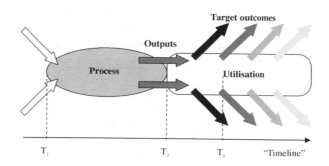

generated every day for as long as the tunnel remains in operation. This would be the case for many projects, but in some instances, utilisation is time limited. Take the case of a five-day exhibition of rare paintings in a provincial city for which the target outcome is "increased awareness of the services provided by a national museum". The bulk of utilisation would take place during the exhibition itself, and thus few, if any additional flows of target outcomes would be expected after the exhibition was closed.

According to the ITO model, a project brings about change when its outputs are utilised by the project's customers. So what form does the utilisation mechanism take? Utilisation represents the total effective operational difference when one processing environment is substituted for another. In the hospital example introduced into Box 2.3, the "Yes" scenario features hospital staff and patients (and possibly others) employing the new outputs. Here utilisation will take the form of certain operational processes being executed (on a day-to-day basis). Some of these processes will be new, some will be modified versions of the old, some may even be the old processes constrained by the new business environment. Because the processes that shape the "Yes" scenario are different from those that shape the "No" scenario, different values will be observed for the waiting time and cost variables. In multiple output, multiple outcome projects (such as the hospital example), many customers will utilise many outputs to generate many outcomes. The linkages here can be displayed with an analytical device called the utilisation map (discussed in Sect. 5.2 in Chap. 5). In such projects, it is possible that certain forms of utilisation by particular customers may actually contribute negatively to target outcomes. For example, the utilisation of (presumably expensive) new medical technology may tend push the cost variable away from its target, and so a trade off may be required between those outcomes that are positively impacted by utilisation of new technology (waiting times) and those that are negatively impacted (operating costs).

2.4.3 Projects and Business Operations

Where does this leave the ITO model of an operational process (as distinct from a project)? The theory behind the ITO model does not require that all processes have their own target outcomes. Two situations can arise. The first is where a process merely contributes to an outcome. In the hospital project introduced in Box 2.3, a task such as "design engagement programme for nursing staff", will contribute to the outcomes of the overall project, but it need not have target outcomes of its own. (Even if it did, we would gain nothing by analysing them). The second is where target outcomes are set for a programme of process executions, but not for individual executions. Take, for example random breath testing by state police. A valid target outcome for this programme would be "reduced incidence of crashes caused by drink driving". Over a year, such an outcome could be measured and compared with some "do nothing" baseline. There would, however, be no point to setting such a target outcome for each execution of the process, that is for each breath test of each driver who was pulled over for testing.

It is the execution of regular operational processes that determines the values of a host of variables that describe each of the three scenarios used to analyse a project's target outcomes (as discussed above).

The following project case provides illustrations of the ITO terminology. A project is executed by the roads authority to improve the quality of line-marking on national roads. A valid ITO model for this project includes:

1. *Inputs*. Funds and labour, measured in dollars and working hours.
2. *Process*. Mark lines. The productivity of this process can be measured in lane kilometres per dollar outlaid.
3. *Output*. Pavement lines, measured in kilometres of fully marked lane.
4. *Utilisation*. Compliant (or non-compliant) behaviour on the part of road users. This particular form of utilisation takes the form of a change in driver behaviour (probably more accurately called "acknowledgement"), measured in kilometres driven within marked lanes.
5. *Outcomes*. Decreased accident rates, measured in serious crashes per year and increased traffic flow, measured in travel time (between two defined points).

The illustration of this project is shown in Fig. 2.7.

If utilisation is going to occur over an extended time-horizon (and this will be the case for only certain projects), should we separate the project environment from the operational environment? And if so how is that separation to be made? Because of its very nature, the project environment is not suitable for ongoing routine business operations and so it is desirable that, at some point, it is replaced with an operational environment. The timing of this can be decided by applying a *test of conclusion* for the project environment. The project environment ends (and the operational environment begins) when *the flow of target outcomes is secured*. Outcomes are secured when any of three conditions is met:

1. The target flow of outcomes is achieved and there is an acceptable probability that it will continue at this level into the future.
2. The flow of desired outcomes is maximised, and despite falling short of target, there is evidence that it will continue at this level into the future.

Fig. 2.7 An ITO model of a road line-marking project

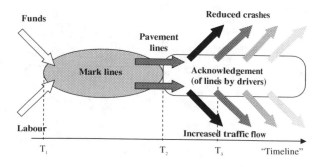

3. The flow of target outcomes is not realised at all and it is expected that it will never be realised.

The project environment can also end in the (pathological) situation where delivery and implementation of outputs does not take place at all (such as when a project is abandoned part way through). In this case, utilisation does not even start.

The formal transition from the project environment to the operational environment is effected with a simple declaration that accountability for target outcomes now passes to an operational business unit. In summary, a number of early instances of utilisation belongs to the project environment, until the flow of target outcomes is secured. From this point on, any remaining flow belongs to the operational environment.

2.4.4 Key Players: Roles, Responsibilities and Accountabilities

It is possible, even at this early point in the exploration of the ITO model to consider not only some key forms of project stakeholding, but also the nucleus of a governance model. Such a discussion requires some additional terms and concepts.

The term "key player" used here simply identifies those stakeholders who play a prominent part in the project. It is not used in a technical sense and requires no definition. Four key players emerge in the immediate discussion:

1. *The funder* (of whom there can be more than one). Who approves the commitment of resources to the project.
2. *The project owner*. Who acts as the funder's agent during execution.
3. *The project manager*. Who "runs" the exercise.
4. *Project customers*. Who, by utilising the project's outputs, generate target outcomes.

The "parts" that key players might play are usefully categorised into three: a role, a responsibility and an accountability, as shown in Fig. 2.8. This figure also confirms that those who play such a part in the project are stakeholders as well (by virtue of the fact that they are commissioned to fill defined roles and so now have an interest in the venture).

Fig. 2.8 The relationship between stakeholding, roles, responsibilities and accountabilities

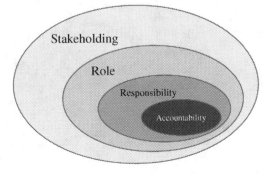

A role for an entity arises when a project-related activity requires the involvement of that entity for its completion. For example, the utilisation of a project's outputs requires the involvement of project customers and so project customers *fill a role* in a project.

A responsibility arises when a role is based on a formal agreement by the relevant entity to participate in certain activities. For example, a project manager is responsible for ensuring that all of the work of a project is appropriately resourced, and so we would expect this arrangement to be reflected in some sort of agreed brief or contract.

An accountability arises when a responsibility is subject to agreed rewards/penalties. An accountability must be accompanied by one or more authorities (powers to take specific actions or make particular decisions). Without authorities, accountabilities collapse into responsibilities. For example, if the project manager is held accountable for delivery of all project outputs (fit-for-purpose, on time and within budget) then he/she must also be granted the authority to deploy the resources made available for the conduct of the exercise. Accountabilities should also be described in a formally agreed instrument of some kind (such as a memorandum of understanding or a contract). It is important to note that, in general, a project customer has no responsibility for (and cannot be held accountable for) utilising the projects outputs.

A funder is an investor in the project who seeks a future flow of desirable outcomes as a return on the funds made available for the exercise. It should be clear that the most important stakeholder in a project is the funder. If he/she does not approve funding (or a continuation of funding), then the project cannot proceed. A funder will only approve a project if the eventual flow of benefits is seen as an adequate return on the funds being invested in the initiative. The term "funder" is defined in the glossary, but for this immediate discussion it should be noted that the funder is the entity who has discretionary authority over the funds required by the project. The funder is not necessarily the "owner" of those funds. While funders may well shoulder accountabilities arising from their *line positions* in their particular organisations, *project* accountabilities stop with funders. Funders are the ultimate stakeholder in the project.

While funders can exist at any level, the approval of an organisation's largest projects tends to fall to the most senior ranks. This frequently gives rise to a practical problem, whereby executives are involved with such a large portfolio that they are unable to play an effective role in the day-to-day leadership of all the projects that they approve. To deal with this problem the role of project owner emerges. A project owner is the person held accountable by the funder for the eventual realisation of a project's business case in general and target outcomes in particular. (It should be noted that the funder can fill the role of project owner if required). Once appointed, the project owner can now commission someone to deliver the project's outputs (fit-for-purpose, within the constraints of an agreed budget and timeframe). Consistent with conventional thinking, the person accountable for delivery of the project's outputs is called the project manager.

A governance model starts to emerge by establishing a client/supplier relationship between the project owner and the project manager.

A project may have more than one funder. This will occur when it is not appropriate for one entity to bear the investment burden alone. The degree of concordance between the objectives of co-funders will shape the project's governance model. At one extreme, all funders may share identical expectations as they relate to objectives and outcomes. Such a project could be approached as if there were only one funder. At the other extreme, funders may have conflicting expectations, incompatible objectives and mutually exclusive target outcomes. Clearly joint funding of such an exercise would make little sense, regardless of how it was governed. Between these two situations are ventures in which the prospective funders find their expectations are compatible, their objectives congruent and their target outcomes distinct but consistent. In such cases, it may well be appropriate to collectively fund a joint project, but allow each funder to appoint a separate project owner. Since an owner is the project manager's client, the project manager now faces the possibility of conflicting advice concerning the direction of the project. To deal with this situation, the governance model provides for a steering committee that is responsible for overseeing the exercise. The steering committee fills the role of client to the project manager. The steering committee must reach a consensus on the instructions and guidance it issues to the project manager.

It is entirely feasible (and quite common) for a particular entity to fill a number of project roles simultaneously. For example, consider a project aimed at reducing binge drinking amongst young women by conducting an alcohol-awareness programme. Because the target community group is presumably better off from a successful project, they qualify as beneficiaries. They are, of course, at the same time the project's customers (in that they utilise the awareness programme to generate the target outcomes) (Box 2.5).

2.5 Illustrating the ITO Methodology: The Project BuyRite Case Study

A case study is used to illustrate various concepts terms, processes and tools that form part of the ITO methodology. "*Project BuyRite*", is based on a real-life exercise conducted by a multinational building products company in the late 1990s. It describes a project aimed at improving the procurement process in a company named *International Concrete Operations Inc.* (ICO).

2.5.1 The Company

International Concrete Operations Inc. (ICO) was a relatively small national concrete operator until it acquired one of its rivals and its two largest suppliers

Box 2.5 ITO-Based Terminology

An Important Distinction: Customers, Beneficiaries and the Client

We have taken a number of steps in our attempt to introduce rigour into the framework introduced here, including:

- Adoption of the ITO model as an organising theory. Much of our discussion emerges as extensions to and deductions from this theory.
- Establishment of a number of principles (which we use to fill a role akin to axioms in other disciplines).
- Assembly of an integrated glossary.

Some elaborating comment is required about the last of these. Selection and adoption of appropriate terms for the elements of our framework involves trade-offs between:

- The need to reflect subtle nuances in the concepts we use.
- Creating new words unnecessarily.
- Acknowledgement of established usage of existing terms.

A particular illustration of the challenges faced here concerns our use of "customer". We use the word in two distinct ways.

- "Project customer" who utilises a project output in such a way that target outcomes are generated. If one were to display project customers in an ITO diagram, they would, of course, appear inside the utilisation rectangle.
- The entity to whom *operational outputs* are delivered (usually involving a commercial transaction of some kind). This use of the word "customer" accords with its commonly accepted meaning. Because the same entity *frequently* fills both roles, some of the existing literature assumes that they are *always* the same. To distinguish these roles we will, where necessary, qualify "customer" with the words "project's ..." or "organisation's".

Other distinctions need to be highlighted as well. Two are particularly important because they are surrounded by considerable confusion in the literature.

- The first concerns the role of project owner. As will be discussed in the section on governance, a project manager is commissioned as supplier (of project outputs) by a project owner. Accordingly, the project owner becomes the project manager's "client". The one entity can be both project client and project customer, but in general, that is not the case. We avoid altogether the term "customer" when referring to the project owner.

> • The second concerns the project beneficiary, defined here as an entity who enjoys a "flow of value" arising from achievement of target outcomes from the project. Beneficiaries can be project customers, but in general, this is not the case. We do not refer to beneficiaries as "customers".

(a quarrying company and a cement manufacturer). An aggressive programme of international acquisitions over the following 15 years saw the creation of one of the world's dominant players in the concrete, cement and quarrying industries.

International Concrete Operations Inc. (ICO) is organised essentially along country lines, with operations in each country being set up as a business unit.

The firm has always valued entrepreneurship very highly, but has tended to lag its rivals in management capability. Weaknesses in a number of areas have left the company exposed to smaller, faster and more agile rivals. The Board has become increasingly concerned about ICO's vulnerabilities, especially in areas where growth has outstripped the capabilities of its management skill pool.

A significant international benchmarking exercise was undertaken to see how well ICO performed in a number of core processes, particularly procurement. This study not only confirmed what the Board had suspected (that ICO was in the bottom quartile for all processes) but also that:

1. The Company has an unenviable reputation amongst major suppliers of being a very poor payer, with 50% of all invoices still outstanding after 90 days. This has had two effects: (1) reliable suppliers are pricing their offers to ICO at a premium and (2) ICO is unable to take advantage of early-payment rebates.
2. Uncoordinated purchasing policies in different business units over many years has led to the growth of large and costly purchasing functions that have not kept pace with modern procurement thinking. Few opportunities for volume discounts are available to ICO.
3. Procurement processes have "grown like topsy", largely the result of adapting local practice as each new company was acquired. These processes are inconsistent, undocumented, inefficient and slow, forcing most business units to maintain unacceptably large inventories to deal with frequent outages. This situation is a significant contributor to the company's very high working capital.

2.5.2 Project BuyRite

The Board acted immediately on the benchmarking report and asked the CEO in each country to undertake a procurement process improvement project.

Charles Edwards, the Australian CEO, created a new position, National Procurement Manager, and appointed Nancy Palmer to fill the role. Nancy Palmer has secured the services of an experienced business process improvement specialist

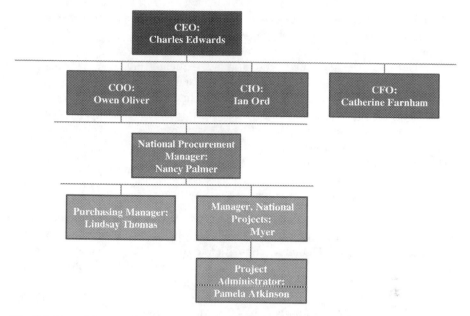

Fig. 2.9 Part of the organisational structure of ICO's Australian operation

(Paul Myer) who will serve as project manager. Paul Myer has just successfully completed another similar project, and has brought with him the project administrator from that exercise, Pamela Atkinson, to fill a similar role on "Project BuyRite".

Preliminary discussions involving the CEO, the COO (Owen Oliver), the CFO (Catherine Farnham), the CIO (Ian Ord), Nancy Palmer and Paul Myer have resulted in the assembly of some initial ideas on the shape of the project.

Early stakeholder analysis identified Lindsay Thomas as an important influencer. Lindsay has been with ICO for 35 years (most of that as a Purchasing/Expediting Manager) and is widely respected for his professionalism.

Figure 2.9 presents those parts of the organisational structure for ICO's Australian operation that are relevant for this illustration.

In Project BuyRite, ICO itself (or, more specifically, its community of shareholders) is the beneficiary of a successful project because it is the Company that experiences the beneficial effects of reduced procurement costs. The procurement staff, who are not shown in this particular chart, are the project's customers because it is anticipated that they will work with the new procurement process, and by doing that, reduce the company's operating costs.

2.5.3   The Emerging Shape of Project BuyRite

In order to specify target outcomes, ICO have addressed the following questions about the proposed exercise:

- What are the dominant characteristics of the current procurement scenario?
- What will happen if we do not intervene?
- What objectives do we seek from intervening in some way?
- Why undertake this particular project at this particular time?
- How does the project fit into ICO's business strategy and priorities?
- What beneficial effects can we generate from intervention?
- What opportunities exist to radically change our procurement-related processes?
- How effective are the best of those?
- What are the downsides of intervention?
- How much will intervention cost?
- How risky is such an initiative?
- What strategy should guide the way we tackle this project?

The background to the case study makes it clear that reduced procurement cost is a desired outcome. Furthermore, because a threshold level of reduction will be set, it becomes a target outcome (although there may be others). Now consider a situation at the end of the project in which it is discovered that, not only have procurement costs been reduced, but also that long-standing frictions between procurement and production have evaporated because stock-outs no longer occur. If reduced friction between these two units was not adopted as a target outcome, then this would be classified as a fortuitous outcome. The question of whether or not fortuitous outcomes are recognised when assessing a project is discussed in Sect. 3.4.

Early discussion of Project BuyRite's outcomes suggested the following outputs:

- (Documents describing) a new procurement process
- An enabling software system
- A panel of preferred suppliers
- A new organisational model for the Procurement Department
- A new office for the Procurement Department
- A performance bonus for Senior procurement staff

According to the ITO model, outcomes are generated when a project's customers utilise outputs. In this case for example, both suppliers and Procurement staff will utilise the panel of preferred suppliers to generate reduced procurement costs. As it happens, later work on the project's scope resulted in significant changes to the outcomes that ICO targeted and also to the list of outputs. More is said about this analysis in Chap. 5.

For each output the work involved will, in due course, have to be described in considerable detail. For example, for the outputs of a new procurement process, the team will need to analyse current procurement practice, examine and rank alternative approaches, select a preferred model, configure enabling systems, train procurement staff in the new process and review how well all this was done. When the work of producing all of the agreed outputs is analysed and described, it will then be possible to estimate the resources necessary to undertake that work.

Two classes of resource will be of particular concern to ICO: those that have to be purchased (such as systems and technology) and the labour of ICO staff who will be assigned to the exercise. These will then become the "inputs" in the ITO model.

2.6 Summary

In this chapter, we have examined four critical features of the project environment by considering: the way it evolves in the course of a project, the elements that give it structure, the engagement of key players and their organisation. We are now in a position to discuss in depth project initiation, the first stage on the life of a project.

Chapter 3
Achieving Success in Projects

3.1 A Framework for Gauging Performance and Judging Success

This section lays the foundations for a comprehensive treatment of project success. The same foundations will also be used in Chap. 5 to explore the processes that surround the decision to approve a proposed project. Although these two topics concern "opposite ends" of a project, they share the same conceptual model.

3.1.1 Perspectives of Success

Baker, Fisher, and Murphy (1988) stated that there is no "absolute" success but only "perceived" success, implying that project success means different things to different stakeholders at different times (Lim & Mohamed, 1999). A project may have many stakeholders whose interests differ widely (and, in some cases whose interests fluctuate wildly over the course of events). (The term "project stakeholder" is covered more fully in Chap. 4, but for the moment can be interpreted to mean "someone with an interest in the project"). The nature of stakeholders' interests shapes their views, not only about the desirability of a proposed initiative, but also about whether or not it finished up "successful".

So whose views about a project really matter? To answer the question we need to distinguish between three related questions (each of which has two parts):

1. "Why is it necessary to make judgements about a business case and (later) about the completed project?"
2. "Whose views about a project *should* be used to approve (or reject) a project and judge its eventual success?"
3. "Whose views about a project *will* decide approval (or rejection) of a project and guide judgements about its eventual success?"

O. Zwikael and J. Smyrk, *Project Management for the Creation of Organisational Value*, DOI: 10.1007/978-1-84996-516-3_3,

As far as the first question is concerned, judgements have to be made about a business case so that a decision can be taken on funding the proposed initiative. Judgements about a completed project, on the other hand, are required for two reasons: to decide if the investment represented by the project was successful and to decide if particular accountabilities surrounding the project have been discharged. The accountabilities in question here are those held by two key players: the project owner and the project manager (as introduced in Chap. 2).

The second question raises normative issues which although important, can only be resolved in social and political forums, the discipline of project management has little (if anything) to add to such discussions.

The third question can, however, be addressed from within the discipline. It is clearly the funder's view that will decide on approval/rejection of a project proposal (because he/she controls the related budget). In a similar line of reasoning, it is the funder's view that determines the eventual success of the resulting investment. Two observations should be made at this point. Firstly, by declaring the funder's views as critical in this way we shed no light on the moral question "*Should* the funder have such a power?". Secondly, there is nothing to prevent the funder from taking the views of other stakeholders into account when making a decision or a judgement about a project. (In fact, the entire stakeholder management process is intended to achieve this in an appropriate way).

However, the concepts introduced in Chap. 2 reveal another two judgements that must be made about a project at its conclusion. At the end of initiation, the funder made a project owner accountable for realising the business case (and achieving its target outcomes). This implies that the funder must eventually make a judgement about the project owner and decide if he/she has successfully discharged that accountability. Likewise, before planning got under way, the project owner made a project manager accountable for delivering the project's outputs in accordance with the business case. This implies that the owner must eventually make a judgement about the project manager and decide if he/she has successfully discharged that accountability.

From this discussion, we conclude that three judgements should be made of a completed project based, respectively, on the performances of the project manager, the project owner and the investment.

3.1.2 The Analysis of Project Performance

A discussion of success requires a robust analytical framework. We propose a structure involving four concepts:

1. *Assessment processes*. To gauge performance. There are two variants which are distinguished according to when they are executed (Campbell & Brown, 2003):

- *Ex ante assessment* (also identified here as "appraisal") is carried out before a project starts, to support a decision about acceptance/rejection of the business case.
- *Ex post assessment* (also identified here as "evaluation") is carried out at the end of a project, in support of judgements about success/failure. (By definition, decisions about success or failure can only be made at the end of a project). Evaluation is based on information provided in a closeout report.

2. *Assessment targets*. To identify the subject of the assessment process. We recognise three targets:

- *The project management process*. Project management success as represented by the project manager's performance.
- *The project ownership process*. Project ownership success as represented by the project owner's performance.
- *The investment*. The success of the investment represented by the completed project.

3. *Assessment tests*. To show how measures of performance relate to a judgement about a project. While *measures* of performance may involve continuous (as well as discrete) variables, by their very nature, *judgements* are based on a classification (which, in this case, is binary). Appraisal results in the classification of a proposed project as suitable/not suitable for funding, while evaluation results in the classification of a completed project as a success/failure. In both cases, the two categories are exhaustive and mutually exclusive, there are no other possible conclusions that can be drawn from each assessment. A test is made up of a specific set of performance variables, a measurement on each of those variables, a set of criteria (taking the form of a reference value for each variable) and a rule showing how to use the resulting measures to make a judgement about the project (See Fig. 3.1).

For example, a test of the project owner might involve:

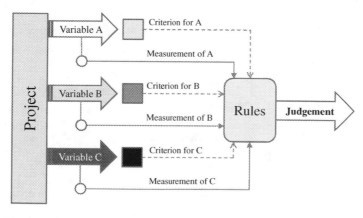

Fig. 3.1 The general structure of success tests in projects

- A variable (such as *worth*).
- A measurement of that variable (such as *the actual value of the worth realised at the end of the project*).
- A criterion (set as *a threshold value of worth*).
- A rule (such as "*If the project's actual value exceeded the notional threshold, then it is judged as successful, otherwise it is judged a failure*"). As it happens, while this illustration is a valid test of project success, it is not a particularly useful one (as will be seen later).

4. *Assessment triggers*. Indicating when a test of project performance should be carried out. There are three triggers—one related to appraisal of a proposed project, the other two related to project evaluation:

- The proposed project should be appraised when the business case is tabled.
- The performance of the project manager should be evaluated when the last of the outputs identified in the project statement of scope has been delivered. In a programme of projects (discussed in Chap. 4), tests of project management success may well be organised for each of the component projects. Similarly, such tests can be carried out on each of a number of selected outputs in a large project.
- The performance of both the project overall and the project owner should be evaluated when the flow of target outcomes has been secured.

The two assessment processes and the three assessment targets suggested above give rise to the six forms of assessment displayed in Table 3.1.

In the assessment framework we propose here, *ex ante* assessment involves absolute versions of the *test* component of the proposed performance framework (described in step #3 above), while *ex post* assessment uses a trade-off variant. In an absolute test, failure of any one variable to meet its agreed threshold is a sufficient condition for a declaration of failure for the candidate being tested. A trade-off test allows for the over performance of some variables to be treated as

Table 3.1 Purpose of the six forms of assessment associated with projects

Assessment target	Who is judged	Assessment process	
		Ex ante assessment appraisal	*Ex post* assessment evaluation
The project management process	The project manager	To decide if the role of project manager is feasible	To gauge the performance of the project manager
The project ownership process	The project owner	To decide if the role of project owner is feasible	To gauge the performance of the project owner
The investment	N/A	To decide if the project should be funded	To gauge the performance of the investment represented by the project

"compensation" for under-performance in other variables. Consequently, a failure by one variables to meet its agreed threshold is not a sufficient condition for a declaration of failure for the project being evaluated.

Sections 3.2, 3.4 and 3.5 discuss these areas of assessment in some detail— with a focus on the three forms of *evaluation* identified in the last column of Table 3.1. Firstly, we cover project management success (because the approach is based on an extension to conventional tests of project performance). Secondly, we consider project success and the way the project owner is judged. Finally, we introduce techniques for evaluating of the investment in the project (which requires some of the concepts used for the other two targets).

3.1.3 Regression Testing

The question now arises "How are we to decide on acceptable and unacceptable trade-offs amongst the variables being used in the (*ex post*) tests of performance?" We propose a regression test, but before describing this, we need to revisit the circumstances under which it will be used. Table 3.1 makes it clear that each of the three forms of *ex post* assessment are carried out following a decision based on a corresponding *ex ante* assessment. For example, overall evaluation of the invest-ment always follows an earlier decision to fund the project, which was, in turn, based on an appraisal (of the business case). So, *ex post* assessment is always undertaken after a judgement has been made about a document tabled earlier.

A regression test uses the structure of the document that informed the relevant *ex ante* decision, but in which the original information is now replaced with the actual data that represents what happened subsequently on the project (rather than what was intended to happen). So if, for example, the original timeframe and cost were 12 months and $300,000, respectively, but the exercise finished up taking 15 months and $400,000, then the higher figures now go into an "achieved" version of the business case (together with actual values of all the other infor-mation that was contained in that earlier document).

More formally, the regression test involves the following steps:

1. In the approved business case, remove the values that were set, anticipated, or estimated for target outcomes, undesirable outcomes, costs and their timings. Remove the stakeholder register, issue register, risk register, schedules, and resource plan.
2. Replace the original elements with those that were actually experienced or realised at the end of the exercise. This gives rise to a new document, which we call the *realised* business case (to distinguish it from the one on which the original funding decision was based, called the *approved* business case). For example, in the ICO case, assume that procurement costs were reduced by 35% (compared with the target outcome of 25%). In the realised business case the outcome would appear with a target of 35% (not 25%) A comment is now

necessary about what happens to the risk register in this process. At the end of the project (when target outcomes have been secured), the only risks that remain are those that apply to any ongoing operational environment created by the project. In Project BuyRite, there would be some ongoing risks relevant to future procurement operations. Residual operational risk is all that appears in the realised business case, all project-related risks have "evaporated" and, accordingly, the residual level of risk exposure will, in general, be significantly less that that reflected in the approved business case.

3. Pose the following question: "Would the original decision-maker (funder, project owner or project manager) have accepted the *realised* business case had it been tabled in support of the original decision?" (Bear in mind that the decision taken by each of these three players is that related to the *ex ante* column in Table 3.1).

4. If the answer to this question is "Yes", the project is judged as successful, if the answer is "No" the project is judged as unsuccessful.

Note that the regression test copes with different project scenarios where some dimensions are more importance than others. Consider, for example, a situation in which the project manager had been authorised to spend more money than initially planned in order to complete the project on time. If the approved business case had shown these (higher) costs, but had been accepted nevertheless, then it would have to be judged as successful.

3.2 Project Management Success

Judgements about the project manager's delivery of the outputs described in the business case are central to evaluation of his/her performance (with possible implications for career progression). It may also, in certain circumstances, provide a basis on which a contractor is remunerated.

3.2.1 The Conventional Treatment

The established tests for project management success are often framed in terms of the 'iron' (or 'golden') triangle (Bourne & Walker, 2004; Gardiner & Stewart, 2000; Jha & Iyer, 2007) in which the committed outputs from the project are delivered subject to three criteria:

1. *Scope/Quality.* In accordance with an agreed specification that identifies all of the deliverables and sets the required level of quality for each (as a list of specified attributes).
2. *Timeframe.* By an agreed date.
3. *Cost.* Within an agreed budget.

In its most common form the iron triangle is proposed as a necessary and sufficient condition for judging success (see Box 3.2 for a discussion of necessary and sufficient conditions for judging success). When used in this way, the three criteria are all applied absolutely and independently, failure of any one criterion implies failure of project management. There is, however, another variant in which trade-offs amongst the three criteria are permitted (Meredith & Mantel, 2009: p. 4). Throughout this chapter, we identify these respectively, as the "absolute" and "trade-off" variants of the conventional test of project management success. Using the model suggested in Fig. 3.1, both are summarised and contrasted in Table 3.2.

Despite their wide acceptance in practice, *both* variants of the iron triangle (absolute and trade-off) fail as tests of project management success. The "digital industrial pressure controller" illustration (Box 3.1) exposes a weakness that arises because the approach ignores *undesirable* outcomes.

Other scholars have also recognised the problems that arise when attempts are made to apply the iron triangle to tests of project success, suggesting that it is partial, unsatisfactory and misleading because it ignores important "soft outcomes", such as the satisfaction of the client or employee development (Baker, Murphy, & Fisher, 1988; Hackman 1987; Scott-Young & Samson, 2008).

These shortcomings are addressed in the next section where we propose an alternative, more general approach.

As a consequence, we conclude that, because it is incomplete, the "iron triangle" is not *sufficient* as a test of project management success and so both tests outlined in Table 3.2 are flawed.

The "trade-off" variant of the conventional test of project management success deals with an inherent limitation found in the absolute variant. Consider, for example, a hypothetical scenario at the end of Project BuyRite, in which Paul Myer, the BuyRite project manager, delivers all of the procurement outputs (process, systems, infrastructure and so on) fit-for-purpose, one week ahead of time but with outlays that exceed an agreed $2.5M budget by only $250 (or 0.01%). This sort of situation is particularly common in projects where outlays

Table 3.2 The two variants of the conventional treatment of project management success

Assessment test component (Fig. 3.1)	Variant	
	Absolute test	Trade-off test
Variables	Scope/quality, timeframe, cost	Scope/quality, timeframe, cost
Measurements	Actual values of each of the three	Actual values of each of the three
Criteria	Thresholds set in the business case	Trade-offs allowed between the criteria
Rules	If threshold for each variable is achieved, then project management is successful If threshold for any variable is not achieved, then project management is unsuccessful	Given delivered scope/quality, if the relative levels of performance against time and cost are acceptable then project management is successful Given delivered scope/quality, if the relative levels of performance against time and cost are unacceptable then project management is unsuccessful

have risen because of conscious decisions to "crash" activities so that timeframes can be reduced. According to the absolute variant of the iron triangle, the management of Project BuyRite would have to be declared a failure. If, however, Nancy Palmer (the project owner) takes a view that the gain of a week was more than adequate compensation for the "loss" of $250, then she may well judge the project management of BuyRite as successful, a not unreasonable conclusion.

The problem here lies in the use of scope/quality, time and cost as three independent criteria, with no trade-offs amongst them permitted. When applied as a test of project management success in this way, the iron triangle collapses to

---

Box 3.1 Illustration: The Digital Industrial Pressure Controller

Why the "Iron Triangle" Fails as a Test of Project Management Success: The Importance of Detrimental Outcomes

Consider a project undertaken by a specialist instrument manufacturer (identified here as "SIM Inc"), in which a new digital industrial pressure controller "D" has been developed under contract for a new client who would otherwise have bought a similar number of a current analogue model "A".

The project manager delivered a prototype of the new instrument in accordance with its specification on time and within budget, but achieved this by putting the team under such pressure that a number of key technical staff resigned towards the end of the project. How is SIM Inc to judge the management of the project (and, by implication, the project manager)? Based on the iron triangle a judgement would have to be made that the project was successful (because all three criteria have been met), but if the loss of staff was considered undesirable, unexpected, avoidable and unacceptable, then *the management of the project* might well be declared unsuccessful, regardless of how the *project itself* was later judged. (We call undesirable end effects that are unexpected, unacceptable and avoidable "detrimental outcomes"). This conclusion implies that the iron triangle fails as a test of project management success, because it ignores those detrimental outcomes that can be attributed to the management of the project.

---

nothing more than a list of three criteria (in which any geometric relationships implied by the triangle are completely ignored). By way of contrast, in the trade-off variant, the geometry of a triangle does have a meaningful interpretation. It can be easily shown that if any one of three variables making up the triangle is held constant, there will, in general, be many combinations of the other two

Box 3.2 Concepts: Necessary and Sufficient Conditions for Judging Success

The Conclusions that can be Drawn from Tests of Success

Not all evaluation tests allow complete judgements to be made about a project. A case in point is the conventional test of success.
  Evaluation tests can be classified as follows:

- *Sufficient test.* If all criteria are met, then the target of the evaluation (the project overall or the project manager) can be judged unambiguously as successful. If, however, the target fails to meet at least one criterion, then no conclusion can be drawn about whether or not it was a failure.
- *Necessary test.* If the target fails to meet at least one criterion, then it must be declared a failure. However, if all criteria are met, then no conclusion can be drawn about whether or not it was a success.
- *Necessary and sufficient test.* If all criteria are met, then the target must be judged as successful. Furthermore, if the target fails to meet at least one criterion, then it must be judged as a failure.

Note that the last test is superior because it cannot lead to the inconclusive result that may arise with the other two. It should also be borne in mind that a situation in which lack of information prevents one from making a judgement about success does not, of course, imply a third category. "Inconclusive" is a characteristic of a *test* not of the test's *subject*.
  Table 3.3 shows how these classes of test link the judgements that can be made about a project to the possible conclusions that can be drawn about whether or not criteria have been met.

Table 3.3 The conclusions that can be drawn from different classes of test

Status of the criteria that make up the test	Type of test used in project evaluation	
	Necessary	Sufficient
All criteria met	Project was not a failure. No conclusion can be reached about its success	Project must be judged as successful
At least one criterion not met	Project must be judged as a failure	Project was not a success. No conclusion can be reached about its failure

variables that are consistent with a feasible project (there are also many combinations of the other two that are infeasible).

Notwithstanding certain desirable features of the trade-off variant, both forms of the conventional approach are flawed as tests not only of both project management success, but (as we shall see) of ownership and investment success as well.

3.2.2 A New Treatment

To rectify shortcomings in the conventional test of project management success, we propose a variant of the regression test. Our discussion has two parts: firstly, we augment the three accepted criteria with a fourth and then we allow for trade-offs amongst them (which is required by a regression test). The four criteria are:

1. *Scope/quality*. An index of delivered scope–obtained by assigning weights to each output and to their actual level of quality.
2. *Schedule overrun*. The ratio of the actual time taken to produce, deliver and implement all project outputs and that indicated by the last approved baseline schedule. The term "baseline" here is a defined word in the lexicon of the project management discipline (effectively meaning "reference"). It should not be confused with two similar expressions used here: *baselining* (to measure the current value of a variable) and a *baseline* document (the business case or project plan).
3. *Cost overrun*. The ratio of the discounted value of project cost (outlay stream), to the last approved (baseline) budget.
4. *Undesirable outcomes attributable to the project manager*. An index based on the ratio between all actual flows of undesirable outcomes (attributable to the project manager) and those anticipated in the last approved business case.

Table 3.4 is a "corrected" version of Table 3.2 showing how the tests of project management success should be constructed.

However, like Table 3.2, this shows two *alternative* tests (absolute and trade-off), so which one is to be applied when making a judgement about a particular project (and its manager)? Because the absolute test is simply a special case of the trade-off test (in which trade-offs have been set to zero), we can conclude that the trade-off test will always apply, but with varying levels of acceptable trade-off. Thus, the last column of Table 3.4 defines a regression test of project management success.

We can conclude, therefore that a necessary and sufficient test of project management success involves four success measures with trade-offs amongst them permitted, and so the ubiquitous triangle is effectively replaced with a tetrahedron, as suggested by Fig. 3.2.

Because the regression test allows trade-offs amongst the four variables, the geometry of the tetrahedron is meaningful. A tetrahedron has four faces, each of

Table 3.4 The "corrected" variants of the two conventional tests of project management success

Assessment test component (Fig. 3.1)	Variant	
	Absolute test	Trade-off test
Variables	Scope/quality, timeframe, cost, detrimental outcomes (attributable to the project manager)	
Measurements	Actual values of each of the four	Actual values of each of the four
Criteria	Thresholds set in the business case	N/A
Rules	If threshold for each variable is achieved, then project management is successful.	Given delivered scope/quality, if the relative levels of performance against time and cost are acceptable then project management is successful
	If threshold for any variable is not achieved, then project management is unsuccessful	Given delivered scope/quality, if the relative levels of performance against time and cost are unacceptable then project management is unsuccessful

which is associated with a fixed value of the (hidden) vertex. For example, the achievement of scope/quality will by driven by manipulation of time, cost and undesirable outcomes.

Table 3.5 illustrates the use of these measures for a hypothetical situation that may arise at the end of Project BuyRite in which (for simplicity of discussion) we will assume that all outputs were delivered fit-for-purpose.

When using the regression test to judge Paul Myer (the Project Manager), Nancy Palmer (the Project Owner) would note that the original committed outputs had been delivered to their agreed specifications and so she is now concerned only with the opposite side of the steel tetrahedron, in the form of a triangle involving cost, time and undesirable outcomes. Beyond this point, the geometric model sheds little, if any light on the question "was Paul Myer successful?", instead Nancy Palmer must make a judgment about whether or not she would have accepted a proposal for a project in which the costs were $500,000 greater, the timeframe 6 months longer, but where only one staff member resigned instead of two.

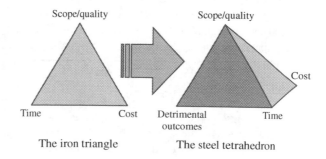

Fig. 3.2 Gauging project management success, from the "iron triangle" to the "steel tetrahedron"

Scope/quality

Scope/quality

Time   Cost

Cost

Detrimental outcomes   Time

The iron triangle

The steel tetrahedron

Table 3.5 A hypothetical scenario at the end of Project BuyRite

Project management success parameter	Relevant values of project parameter	
	Targets established in business case (as thresholds)	Actual values achieved at end of project
Scope/quality	Set in a scoping statement	Approved scope delivered
Cost	$2.5M	$3M
Timeframe (months)	30	24
Undesirable outcomes	Loss of two key staff	Loss of one key staff member

3.2.3 Judging Project Management Success

Care must be exercised when applying the suggested four project management success measures because there are occasions when criteria appear not to be met, but where such an observation has no implications for project management performance, as suggested by the following three examples:

- *Time overrun.* Extraordinary demands for heavy lift marine cranes in China slow the rate of delivery of pre-cast sections of a new concrete box girder bridge in Malaysia, causing significant delays in the opening of a new motorway. If such an event was judged unforeseeable, then the project manager could not be held responsible for the failure to meet the schedule. If, on the other hand, there had been considerable discussion about possible accelerated investment in Chinese port infrastructure, then a view might well be taken that such a threat should have been recorded in the bridge project risk register and that the project manager failed by not recognising the risk.
- *Detrimental outcomes.* Only those that can reasonably be treated as "caused" by the project manager are relevant to judgements of project management success. If, in the case of the digital pressure controller (Box 3.1) it is discovered that, in addition to the anticipated cannibalisation of the old analogue product "A", sales of another multi-function product "M" were significantly reduced as well, then one would be hard pressed to "blame" the project manager. Cannibalisation of existing products (whether expected or not) is certainly a detrimental outcome, but one that has nothing to do with the management of the project.
- *Undesirable outcomes that have been identified in the business case.* Project management success should not be affected by undesirable outcomes that have already been identified and anticipated in the business case. Table 3.6 displays the impact on project management success arising from various combinations of both the type of undesirable outcome and the role of the project manager in its occurrence.

As can be seen in Table 3.6, project management performance is impacted by those detrimental outcomes that are unambiguously attributable to the project manager's actions. Those that have arise for other reasons are ignored when considering the success of the project manager (although, as will be discussed

Table 3.6 Impact of undesirable outcomes on project management success

Actual level of undesirable outcomes, compared to the business case	Is the difference attributable to the project manager?	
	Yes	No
Greater	Negative impact↓	No impact↔
As expected	No impact↔	No impact↔
Less than	Positive impact↑	No impact↔
New (not identified in the business case)	Negative impact↓	No impact↔

later) they are taken into account when measuring project ownership and investment success.

There is a practical issue to be considered when applying a test of project management success. In the case of the steel tetrahedron, what thresholds are to be used as reference levels when making trade-offs amongst all of the four variables? While at first this seems obvious (use the values set in the business case), the situation becomes less clear when one considers that all four parameters (scope/quality, time, cost and detrimental outcomes) will be the subject of continuous revision from the point when the original business case is accepted through to the

---

Box 3.3 Case study: The School Refurbishment Programme

Why Increases in Timeframe and Cost do not Necessarily Imply a Project "Blowout"

Consider the following newspaper headline (based loosely on a real case from the mid 2000s)—"Massive blowout on schools refurbishment programme". The article then goes onto declare that "the government's new initiative to improve facilities at state schools is already in trouble only 6 months into the anticipated 2 year project because early surveys have revealed that most buildings slated for refurbishment in a particular region have asbestos roofing which must be removed by specialist contractors. It is understood that this will delay the project by 4 months".

If the project is indeed delayed, must we infer that the exercise has failed its time criterion? Not necessarily. If the project risk register had not only identified the "presence of asbestos" as a threat, but had also predicted a 4 month delay in removal (together with a corresponding cost increase), then the project timeframe and cost constraints have not been violated. Why? Because this has been accepted in the business case as a possible scenario, and it has been funded accordingly. If, on the other hand, such a situation had not been identified, then the timeframe and budget have certainly been "blown", but even this does not imply a failure of project management, unless it can be shown that such a risk should have been identified.

last revised plan. The most pragmatic approach appears to be one in which the last approved baseline document provides the thresholds required in the eventual test.

3.2.4 Project Management Success in Practice

The purpose of this section is to examine the extent to which organisations meet project management success criteria. Projects that did not meet these targets are well known. For example, projects that exceeded their approved budget include the Channel Tunnel (80%), the Brooklyn Bridge (200%), Denver International Airport (200%), the Boston Big Dig (275%) and The Sydney Opera House (1,400%) (Flyvbjerg 2007). On the other hand, some projects have been completed in time, for example, New York's Empire State Building, the Paris' Eiffel Tower and the Guggenheim Bilbao Museum in Spain (Flyvbjerg, 2005).

Appendix C presents the results of a study conducted by the authors in which project success was analysed in 776 projects across seven industries in different countries. Table 3.7 compares success rates for three of the four project management success measures discussed above. (Undesirable outcomes were not included in this analysis). The table shows the mean values in each case.

If evaluated using the absolute variant of the conventional test for project management success (involving separate criteria for scope/quality, time and cost), then this table suggests "poor" results for most projects (acknowledging, however there are significant differences between industries). Given our earlier criticisms of such a test, it is necessary to qualify this conclusion. Firstly, it does not account for detrimental outcomes. To the extent that they have been realised in individual projects, then the "true" rates of failure would be *at least as great* as indicated here. Secondly, this study is not based on a regression test for each of the projects included in the sample (it uses absolute criteria where trade-offs have not been accommodated). A regression test would indicate "true" rates of failure *no greater than* indicated here. On the assumption that detrimental outcomes and trade-offs

Table 3.7 Project management success levels across industries

Industry type Scale Goal	N	Schedule overrun % Minimise	Cost Overrun % Minimise	Quality of outputs 1–10 Maximise
Engineering	98	13.7	12.8	8.4
Software	237	19.8	15.8	7.1
Production/maintenance	67	22.7	17.1	7.0
Construction	23	6.0	5.3	8.3
Telecommunications	93	17.0	11.5	8.0
Services	64	12.2	12.0	7.8
Government	154	23.2	15.3	7.7

are no more common in one sector than another, we can conclude that the *relative performances* of the sectors shown in this table are valid.

The performance of construction and engineering organisations across all project management success dimensions is worthy of note. A number of factors may explain this result. Construction and engineering projects are subject to particularly thorough planning, analysis and review. This preparation is often concerned with an examination of numerous options before a preferred approach is finalised. While scope can and does change in engineering and construction, anecdotal evidence suggests that, because scope is normally well defined at the outset, it rarely has to change in response to scoping *errors*. When it does occur, scope change is treated very seriously and managed in a thorough, systematic and formal way.

A number of explanations can be advanced for this high level of capability across the industry.

- This is a mature discipline which can trace its lineage back for hundreds (if not thousands) of years.
- Competition amongst firms in the sector is relatively intense (at least across the countries covered by the survey) forcing participants to maintain high standards of performance.
- Construction methodologies tend to change relatively slowly (possibly a by-product of their maturity). This allows firms time to improve their performance from project to project.
- Project managers are, in general, competent in a number of core skill areas—especially those associated with coordination and integration. This capability is developed on-the-job in, what amounts to, an apprenticeship involving participation as a member of team on a number of significant projects, before taking on full responsibility for a major project. This programme of professional development will often include a period of mentoring by a senior project manager.

By way of contrast, the software sector for example, achieves relatively low scores. This field is characterised by significant technological uncertainty, rapid change, and high staff turnover. Projects are subject to frequent changes of scope (often to deal with problems arising from poor initial scoping). Another factor may explain much of the relatively poor performance of this sector when compared with engineering and construction. Software development appears to be characterised by labour-intensive "hand-crafting" processes, each of which tends to involve a high level of novelty. In engineering and construction work, many outputs emerge from assembly-type processes in which components are bought off-the-shelf and "put together" (possibly after modification). Not only are the costs of the components known, but also the assembly processes are relatively standardised. If this conjecture is correct, then the ratio of labour/output-value for projects undertaken by the software industry will be significantly higher than for the engineering/construction sector. Estimation of outlays for purchased components involves only counting and pricing. Estimation of labour involves judgements about work. Because the reliability of labour estimates is inversely related to

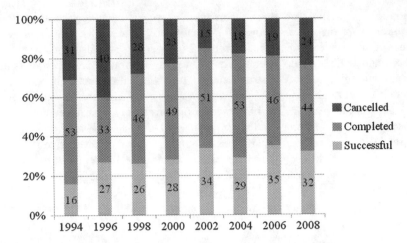

Fig. 3.3 Project management success in the information technology sector

the novelty of the work being estimated, we can conclude that, in general, project estimates for the software sector will be less reliable than those for engineering/construction. Interesting evidence for the low rates of project management success in this sector can be found in the bi-annual report conducted by Standish Group, known also as the "Chaos report". Figure 3.3 presents the results of this survey since 1994.

The Standish Study also reported generally low levels of project management performance in the information technology (IT) sector, with about third of all projects surveyed being cancelled (that is, no outputs were delivered) and another half completed but judged as unsuccessful. The remainder were judged successful if completed on time and within budget, but with no reference to their quality or undesirable outcomes, leaving open the possibility that project management success rates were actually much lower than reported. It is difficult to draw any useful inferences from these results for levels of project success in the IT sector (for which there is a dearth of empirical evidence). Two lines of *a priori* reasoning lead, unfortunately, to opposite conclusions. On one hand, one might argue that while formal business case regression testing may reveal higher levels of success with projects than with their management, the monotonous procession year-by-year of studies reporting extraordinary rates of failure offer little hope that such a situation actually prevails. On the other hand, we see a long history of ongoing investment by organisations in new IS/IT-related projects. Presumably, senior management is making considered judgements to proceed with these ventures and is, therefore, being encouraged by acceptable "true" underlying rates of success.

Although criticism has been levelled at the methodology used in the Standish Group's study (Jorgensen & Molokken-Ostvold, 2006), the general perception of projects undertaken in the IT sector is that their success rates are low. This view is supported by similar results from other studies. For example, an investigation of business projects with significant IT outputs undertaken by mortgage firm

(Tichy & Bascom, 2008) revealed that 11% were cancelled, 78% challenged (completed but unsuccessful) and only 11% successful. Poor management is considered the major reason for these disappointing results in the IT sector. Specific areas of weakness included leadership, stakeholder management, estimation methods, risk management and management support (McManus, 2004).

In summary, while numbers may vary by industries and type, most projects do not achieve the criteria for project management success.

3.3 The Worth of a Project

Discussion about the other two areas of evaluation (project ownership and investment success) requires some additional concepts and terminology.

The decision to fund a project involves ranking it with all other alternative proposals. Projects are ranked with an overall qualitative index of their "attractiveness" for investment. The attractiveness of a project is determined by two other qualitative variables: "worth" and "riskiness".

The worth of a project is derived from benefits, disbenefits and costs, according to the general equation 3.1, but (because "worth" is merely an index based on variables that do not share common units of measurement), the conceptual function linking these components will be peculiar to each project.

$$\text{Worth} = \text{Fn}(\text{Benefits}; \text{Disbenefits}; \text{Costs}) \tag{3.1}$$

Benefits and disbenefits are, in turn, related to outcomes. An exploration of the relationships among the three measures is required before discussing worth in more detail. Worth is a qualitative measurement of the overall net value (as perceived by the funder) that is generated by an initiative. At the end of a project, for example, we would like to know about the value created from our investment. Ideally (in that case) worth would be measured in dollars, but (for a number of reasons) this is often not possible. The following sections discuss project worth and its components in detail.

3.3.1 Evaluating a Project's Worth

The outcomes from a project result in "flows of value" to (or from) certain stakeholders. Consider an initiative by a state health department to improve the performance of hospitals in a particular region by re-engineering administrative processes and medical procedures. A target outcome "reduced waiting time for elective surgery" is set for the project. Achievement of this outcome would give rise to a benefit for every future elective surgery patient, each of whom would value the effect in some way. In general, projects will also generate undesirable outcomes which (through a similar mechanism) give rise to "flows of value" away from certain other stakeholders. For example, staff may be faced with a period of disruption and radical change in the way they perform their duties. To rank this initiative amongst others, the funder (the state government in this case) has to ascribe a single value to its worth by accounting not only for all these different "flows of value" to (or from) different stakeholders, but also for the cost of the exercise. In summary, the worth of a project attempts to account for: one or more outcomes (both desirable and undesirable), each of which will give rise to multiple flows of value (as both benefits and disbenefits) to multiple stakeholders and all of resources that involve some sort of outlay (or cost).

Conceptually then, the *worth* of a project to a funder is obtained from a function of three classes of variable:

1. *The benefits generated by the project*: related to the value of desirable outcomes as perceived by the funder.
2. *The disbenefits arising from the project*: related to the value of undesirable outcomes as perceived by the funder.
3. *The project's cost*: related to the cost of developing and (later) maintaining the project's outputs.

All three variables will typically take the form of a flow (over time) and, as a result, each could be displayed as a set of graphs or bar charts. Because, the measures associated with all the variables are usually incompatible, a mechanism has to be found that will allow worth to be determined in a meaningful way. Numerous approaches have been proposed to address this problem, including, for example, economic cost–benefit analysis, multiple-criteria analysis and making trade-offs amongst the variables.

3.3.2 Measuring a Project's Worth: The Limitations of Financial Units

The practice of assigning arbitrary monetary value (such as dollars) to any non-dollar-valued target outcomes bears further discussion because it sometimes masks the problem of incompatible measures without solving it. Six arguments can be put forward in support of this claim:

1. *Equivalence.* A set of dollar-valued effects is not *equivalent to* the non-dollar-valued variable from which they derive. Take the case of a project to reduce the incidence of domestic violence in a particular community by 1,000 cases/year. Assume that it has been decided to assign a dollar value to each case using some sort of economic model and that an analysis reveals a shortfall compared with the cost of the initiative. Is that the end of the discussion, or could a case still be mounted in support of the initiative? If we allow a trade-off between the way the funder values the reduced number of domestic violence cases and project cost, then a decision could very well be taken in favour of funding, despite the apparent negative dollar worth.

2. *Conversion.* The application of a dollar conversion factor does not convert a non-dollar variable into a real flow of money for the funder. (Whether or not a project *should* generate financial inflows for the funder is an entirely separate issue).

3. *Measurement.* As a gauge of eventual success, measurement of a dollar surrogate is not equivalent to measuring the target outcome. Again, using the example of the domestic violence initiative, what variable will be measured to judge success (*ex post*)? There appear to be only two options: the actual number of incidents of domestic violence or the actual value of the dollar surrogate. However, if the latter approach is used, how is it to be calculated? Since the surrogate is obtained as the product of the count of cases and the assumed dollar value of each case, then success will be sensitive only to the count of cases, the dollar value of each case can have no bearing on the issue.

4. *Pooling.* The fourth point about the limitations of dollar-valued variables requires some preliminary discussion of the concept of "pooling". Consider a project in which some of the benefits and some of the disbenefits happen to take the form of "natural" flows of money. In general, the costs of all projects take the form of money flows. Depending on the particular circumstances, various subsets of these flows can be pooled by particular stakeholders. For example, an initiative to boost tourist visits to a historic town is to be funded jointly by a state tourism authority and by local businesses with the target outcome of "increased sales by local businesses". A variety of scenarios can be postulated for this project, each based on the possibilities of pooling the money flows that it generates, and each giving rise to its own peculiar treatment of the worth of the venture. Compare scenario "Y" (in which each firm looks after its own sales, costs and project levies without any sharing of any moneys) and scenario "Z" (in which all moneys from all sales, costs and levies on all firms flow through a single bank account administered by the state tourism authority and eventually disbursed according to some agreed formula). Clearly the behaviours of the participants will be heavily influenced by the existence of a pool (as in scenario "Z") or otherwise (as in scenario "Y").

5. *Valuation.* Regardless of whether an outcome is measured in dollar or non-dollar terms, its benefit must be gauged from the funder's perspective, not the beneficiary's. Consider a project funded by a city council to attract a major international power boat race, with target outcomes of increased visitations by tourists to the city and increased volumes of sales by local tourist operators. Assume that the project generates significant revenue for an airline that operates scheduled flights to the city, but which is based elsewhere and does no significant business locally. The (non-target) outcome of increased contribution margin (gross profit) to the airline certainly has a dollar value, but in terms of benefits, this outcome might well have zero value to the funding council.

6. *Empirical validity.* If a non-dollar-valued target outcome is given a dollar value by applying a formula, the coefficients in that formula must be capable of validation, whereby their value can be tested empirically as correct or incorrect. Take the scenario in which, for the domestic violence initiative, a dollar value has been assigned to each case using a formula derived from an economic model. Because

the financial valuation of the project is sensitive to both the number of cases and the value assigned to each, the question arises "How do we know if the dollar value being used is correct?" Clearly if it is incorrect, then the calculated financial worth of the project will be wrong. If an assigned or assumed value in a conversion formula cannot be empirically validated, then the formula cannot be used in the analysis of the project.

There is a special class of project in which it is possible (and desirable) to simplify the collections of variables that represent a project's benefits, disbenefits and costs. This occurs when two conditions are met: that the "natural" unit of measure for particular variables is dollars (in other words some variables take the form of a real flow of money) and that the funder is able to pool all these flows. In those cases, all the dollar-valued variables (including relevant benefits, disbenefits, project costs, and the ongoing marginal costs of operating the project's outputs) can be "compressed" into a single quantity known as net present value (NPV). NPV uses discounting, which is the process of converting a future flow of money to a single numeric quantity called its Present Value.

Box 3.4 Illustration: Improving the Safety Record of a Construction Company

Trading-off Net Present Value against Non-Dollar Target Outcomes

Consider a project by a construction company to institute a major industrial safety campaign. Currently, the organisation has poor safety practices, resulting in high insurance premiums and an unenviable accident record. The company also has (what is in the eyes of the Board) an unacceptable ranking in an annual award for excellence conducted by the professional association of builders to which the firm belongs.

Two benefits are sought from the initiative: reduced insurance costs and increased ranking in the excellence awards. The first has been specified as an annual reduction of $250,000 and the second as a move in industry rankings, from the 90th percentile to the 10th percentile. The project will involve an initial outlay of $1M with ongoing costs of $185,000. Because the two dollar-valued variables (benefit and cost) will be automatically pooled by the construction company, it is meaningful to combine them into an NPV of the project. At a 10% discount rate (the assumed market cost of money), the project has an NPV of −$350,000. In other words, the notional value of the dollar-valued target outcome falls short of the projects costs by $350,000. The decision on going ahead now involves a (relatively simple) trade-off between the "loss" of $350,000 and the firm's improved ranking on the industry ladder.

The following example exposes a situation where, although three flows have dollar measure, they cannot be used in an NPV calculation, because only two of them can be pooled. In project BuyRite, three variables could be given unambiguous dollar measures:

1. ICO's reduction in procurement costs
2. ICO's outlays on project resources
3. The reduced working capital of preferred suppliers (arising from quicker settlement of invoices issued to ICO).

The first two of these could be combined into the project's NPV, but the third could not because ICO is unable to draw it into a single financial pool. (In other words, no-one is able to "net off" the three effects).

It is clear that a funder will be satisfied with achievement of some minimum level of worth for the project or better. Accordingly, during appraisal, a decision to fund will require that the *anticipated* worth of the project to the funder is no less than this minimum. Consequently, in a business case the targets set for the three components of worth are, in reality, *thresholds* (a minimum required level for the benefits and a maximum for each of disbenefits and costs). Because appraisal is concerned with a prediction of worth, the reliability of those predictions must be taken into account. Once a business case is accepted, the project's management focus is on achieving the threshold established for the project's worth.

Why do we express our project goals as thresholds, rather than as optima? For example, why would we set a target (of, say, 25%) for reduced procurement costs in Project BuyRite, rather than set a goal of "minimised procurement costs"? The answer is found by considering the tests that are implied by each goal. The 25% threshold is simply tested by comparing it against actual performance. By way of contrast, there is no test that can confirm the achievement of "minimised" procurement costs. Behaviour that supports the achievement of an acceptable threshold is called "satisficing" (Simon, 1976), to distinguish it from "optimising", behaviour that supports achievement of a maximum or minimum.

A discussion now follows of the three components that determine the project's worth—benefits, disbenefits and cost.

3.3.3   A Project's Benefits

Benefits are driven by target outcomes. Benefits take the form of a flow of value from the perspective of the funder to an entity arising from the generation of desirable outcomes. In general, the relationship between outcomes and benefits is many-to-many. Some target outcomes take the form of benefits as they stand, while others represent flows of identified, but undefined, benefits. In Project BuyRite, the target outcome "Reduced costs of procurement" translates directly into a (financial in this case) benefit to ICO. "Reduced payment times to suppliers" on the other hand certainly qualifies as a target outcome, but (as it stands) is not expressed in the form of a benefit to ICO. The entity experiencing a gain of value is called a beneficiary of the project. The benefits of a project need not flow to the funder, nor need they have a dollar measure. Take, for example, a project by a charitable institution to reduce the numbers of children sleeping on the streets—that the benefits flow to the children (rather than the charity) has no bearing on a funding decision.

The foundation work on a project includes setting target outcomes. The assessment of worth then involves judgements about the values of the project's target outcomes.

A particular flow of value to a particular entity will be gauged differently by the beneficiary and the funder. For example, reduced settlement times for supplier invoices are sought as a target outcome in Project BuyRite. Presumably, this effect would be valued very highly by ICO's panel of preferred suppliers because it would translate directly into a financial gain for each of them. The value to a supplier could be found as the NPV of the expected change in the timing of receipts from ICO. For ICO this is certainly important as an outcome because it provides an incentive for suppliers to seek membership of the panel. Its value to ICO as an outcome, however, is unrelated to the value seen by suppliers (amongst other reasons, their financial gain is, in fact, ICO's financial loss). While suppliers see a financial benefit, ICO sees a strategic benefit.

So whose view of benefits is relevant to a project's "intrinsic worth"? The answer is "The funder's", because it is the funder (alone) who will make a decision on the project's future.

3.3.4 A Project's Disbenefits

Disbenefits are driven by expected undesirable outcomes. Disbenefits take the form of a loss of value by an entity arising from the project. The entity experiencing the loss of value is called a disbeneficiary of the project. The disbenefits term in the equation of worth captures not only those detrimental outcomes recognised in the new test of project management success, but also all other undesirable outcomes as well. Mirroring the above discussion about benefits, particular flows of value away from a particular entity will be valued differently by the funder and the disbeneficiary. "Whose valuation of each disbenefit is the one to be used when determining the worth of a project?" Again, the answer is "The funder's", for the same reason, that if the funder does not fund, the project does not go ahead.

3.3.5 A Project's Cost

It is common to find the concepts of costs and disbenefits being used inter-changeably. Here we define them separately and use them in distinct ways. Costs

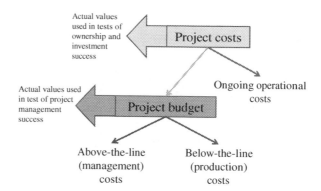

Fig. 3.4 The relationships amongst various project cost elements

are driven by the resources needed to deliver the project's committed outputs, while disbenefits arise from undesirable outcomes. Project costs have three components:

1. *Production cost.* (otherwise known as "below-the-line" cost) arising from the acquisition of the resources required to produce the outputs identified in the project statement of scope.
2. *Management cost.* (otherwise known as "above-the-line" cost) arising from the project's planning, management and administration.
3. *Operations cost.* associated with any (marginal) outlays required to support the ongoing operation of any of the project's outputs.

For example, a transport company undertakes a project to increase customer satisfaction by allowing online consignment tracking. When calculating the costs that will be used to gauge the project's worth, the company will have to recognise not only the outlays to develop the online tracking facility, but also the ongoing costs incurred in operating and maintaining that facility.

While all three of the project cost components must be taken into account when analysing the project's financial performance (using, for example NPV), only the first two (production and management cost) are recognised in the project budget, for which the project manager is accountable. Figure 3.4 summarises the classification and treatment of these various project-related costs.

Identification and valuation of production and management costs is relatively straight-forward, however operations costs present a number of analytical challenges that are discussed in Chap. 6.

3.3.6 Calculating a Project's Worth

The following example illustrates the way in which a project's worth is determined when all significant variables have simple dollar values.

A pharmaceutical company has estimated the following parameters (in which discounting is ignored to simplify the discussion) for a project to manufacture a new drug:

- Product life: 5 years.
- Benefits: Target annual sales = $400M (or $2,000M over 5 years).
- Disbenefits: Lost revenue from cannibalisation of an existing product = $30M a year (or $150M over 5 years).
- Cost: Total outlays attributable to venture. This is made up of two components: outlays to acquire the resources demanded by all the work of the project and outlays required by the resulting manufacturing operations. Project costs can, in turn, be broken down into two sorts: below-the-line and above-the-line. Assume that below-the-line costs = $500M, above-the-line costs = $40M and annual operations costs (to produce the new product) = $25M per year. ($665M all up).
- The project's worth is then:
  Worth = Benefits – Disbenefits – Costs: $1,185M = $(2,000 – 150 – 665)

While the mathematical treatment in this case is the same for disbenefits and costs (whereby both are subtracted from the benefit), the loss of potential revenue qualifies as a disbenefit (not as a cost). This distinction becomes critically important to the project manager who will be held accountable only for meeting a budget constraint of $540M (=$500M + $40M). He/she would not, in general, be held accountable for limiting the product cannibalisation to $20M, nor for achieving the target future annual production cost of $25M.

3.3.7 Benefit–Cost Analysis and ITO-Based Project Assessment

Economists know a lot about investment in projects. This is reflected not only in the formidable suite of decision-making tools that have been assembled over many years, but also in a particularly robust theoretical framework that underpins those tools. Benefit cost analysis identifies a class of technique, which seeks to answer the question "How does the value of beneficial effects from the project on one hand compare with the value of the resources it consumes and any damaging effects on the other". The answer does not take the form of a decision but is instead critical information that should be used when making a decision about a proposed initiative. A complicating factor arises from the fact that this question will have different answers depending on whose view is taken of benefits, resources and disbenefits. While the most sophisticated approaches to benefit cost analysis have been developed for public sector investment, variants have also been adopted widely in the private sector.

From this we can conclude that the respective roles of benefit cost analysis and ITO-based project assessment are similar. Differences, however, lie in the detail of the analytical techniques that apply in each case. Table 3.8 suggests some of the more important characteristics that distinguish ITO-based project assessment

Table 3.8 A comparison of benefit cost analysis and project assessment

Characteristic	Conventional benefit–cost analysis	ITO-based project assessment
Benefit valuation	List all *outputs*	List all desirable *outcomes*
	Assign a notional price to each *output*	Assign a notional (judged) index of value to each desirable (non-dollar) *outcome*
	The dollar benefit (for each period) is calculated as the sum of all flows of *outputs* * notional price of each	The index of benefits (for each period) is calculated as the sum of all flows of desirable *outcomes* * notional value of each
Disbenefit valuation	Disbenefits treated as a special case of "cost"	–
	List all undesirable outcomes	List all undesirable (non-dollar) outcomes
	Assign a notional market value to each undesirable outcome	Assign a notional (judged) index of value to each undesirable (non-dollar) outcome
	The dollar disbenefit (for each period) is calculated as the sum of all flows of undesirable outcomes * notional value of each	The index of disbenefits (for each period) is calculated as the sum of all flows of undesirable outcomes * notional value of each
Cost valuation	List all resources demanded by project.	Calculate opportunity cost of each resource.
	Dollar cost (of resources) = Sum of all	(opportunity costs of all resources)
Value the project	Project valued in dollars	Project valued as an arbitrary index
	The value of the project is the sum of the present values derived from each of three flows: benefits, "disbenefits" and costs	The value of the project is its worth, obtained as a function of (benefits, disbenefits and costs). Benefits and disbenefits are each made up of lists of variables with inconsistent units. Cost, on the other hand is always measured in dollars. To accommodate inconsistent units, worth must be derived using any of a number of particular techniques, such as multi-criterion decision analysis or by making trade-offs amongst the variables

(as discussed throughout this chapter) and a commonly accepted variant of benefit cost analysis (Campbell & Brown, 2003).

A comment is necessary about the treatment of financial variables in ITO-based assessment. Costs can always be given an unambiguous dollar measurement, regardless of the form of analysis being undertaken, and so financial variables will appear in every attempt at project valuation. In addition to costs, other variables may emerge that have "natural" dollar units, for example the target outcome in Project BuyRite "reduced procurement costs". In those cases, the treatment of benefits, disbenefits (and, of course, project costs) should each include a step to calculate the Present Value of all the associated financial variables. This means that the analysis of benefits, for example, will result in a catalogue of the flows of all desirable outcomes, including a single present value figure representing all of the dollar-valued benefits. Similarly, the analysis of disbenefits will result in a catalogue of the flows of all undesirable outcomes, including a single present value figure representing all of the dollar-valued disbenefits. Finally, the analysis of costs will result in a single present value figure representing all of projects resources. When calculating worth, all three of these present values will be netted-off before then being incorporated (as a single numeric value) into the formula adopted to compute worth.

3.4 Project Ownership Success

Just as the project manager is evaluated, so too is the project owner (because both hold an accountability). Whereas judgements about the performance of the project manager are based on only some of the parameters that appear in the last approved business case, the success of the project owner is based on the entire business case.

3.4.1 Judging Project Ownership Success

When a project is approved, the funder is essentially saying "I am making an investment in the project defined and described by the business case, which I expect to see realised". At that point, he/she commissions a project owner to act as his agent and makes that person accountable for realising the business case in general and the project's target outcomes in particular. Eventually a judgement will have to be made about how successfully the project owner has discharged that accountability.

The business case provides the foundation for judging the project and the performance of the project owner. A naïve test would simply ask if the (intended) business case had been achieved (or exceeded). According to that approach, if the actual values achieved for every critical project parameter (target outcomes, costs, timeframes and so on) were no worse than the targets set for those same parameters in the business case, then the project owner would be judged successful (as would the project as an investment). However, such a test does not allow for acceptable trade-offs amongst key parameters (such as outcomes, costs and risk) and hence it must be modified. Consider just two variables from a business case in which worth is set at 100 (arbitrary units) and risk is set at 0.1. (Normally, project risk cannot be gauged as a probability in this way). At the end of the project assume that all risk has "evaporated" to zero, and the achieved worth is 95. Assume, furthermore, that all of the parameters in the business case have been achieved exactly, except for benefits (which fell 5 units short of their target). When judging the project owner, the funder would have to accept that the achieved business case in fact *exceeded* the approved business case because a riskless 95 units of worth has a greater intrinsic value that 100 units with a risk of 0.1 (which has an expected value of only 90 units). From this discussion, we confirm that, while the project owner will be judged on achievement of the business case, a regression test must be used (which allows for trade-off amongst the parameters that had been set). The criteria used in the test of success for the project manager are a subset of those used in the test of success for the project owner. While a project owner has to achieve the equivalent of the entire approved business case to be successful, the project manager has only to achieve part of it. (For example, target outcomes are not considered when judging the project manager). Despite this relationship, results from one of the two tests have no implications for the other. In other words, the judgments made about the project manager and project owner at the end of a project can result in any combination of success/failure for the two key players.

3.4.2 The Regression Test of Project Ownership Success

We propose a variant of a regression test for judging the success of the project owner based on the approved business case as displayed in Fig. 3.5 in which:

- Project "I" represents the approved business case. This was approved by the funder and provided the basis on which the project owner was appointed.
- The lower curve (the "project investment frontier") defines regions of investment success and failure (discussed in Subsect. 3.5.2).
- The upper curve (a "business case value contour") shows all those combinations of worth and risk that the funder regards as being equivalent to "I". The contour is peculiar to each business case. In general, the curve will intersect the Y-axis at a value below that for the point "I", and rise monotonically through "I". The values of worth recognised when establishing this contour are adjusted to account for relevant detrimental and fortuitous outcomes.

Fig. 3.5 The regression test of ownership success

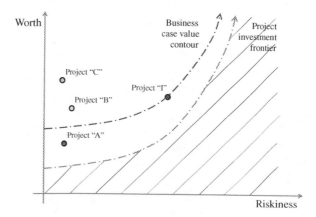

- The region above the *business case value contour* defines those achieved business cases on which the project owner would be judged by the funder as successful.
- The region between the business case value contour and the project investment frontier defines the region where a completed project would be considered a project (ownership) failure, but an investment success.
- "A", "B" and "C" all represent different scenarios for I when completed. Under scenario C, the project finished up with greater worth and lower risk than established in the business case (significantly more attractive than I). Under scenario B, despite its lower worth, the project finished up more attractive than anticipated (because it is above the business case value contour). Both B and C would be judged as successes for the project owner (and, by implication, as investment successes as well). Under scenario A, the project would be judged a failure for the project owner, but again an investment success.

Consider the illustration of the new digital industrial pressure controller (introduced in Box 3.1). Table 3.9 shows both the anticipated and actual (magnitude and timing) of target outcomes, undesirable outcomes and costs (as they would appear in a business case). For simplicity of discussion, this table ignores the time-value of money. It also assumes that each of these is a *one-shot* figure, so there is only one sale of the product to the client, and that the project is riskless.

It should be noted that, because the target outcome is expressed as "gross profit", ongoing operational costs are already taken into account in both scenarios and so the only costs relevant to the example are those incurred from the project's above- and below-the-line work. According to the approved business case (summarised in the "A" scenario), the net financial worth from this project should have been $4M, calculated as the value of target outcomes, less the cost of undesirable outcomes and less the cost of the project itself ($10 - 2 - 4 = 4$). It is clear that although the project was intended to generate a surplus of $4M, it actually generated only $3M ($11 - 2.5 - 5.5$). If the approved business case was achieved (that is the actual figures finished up being the same as those in the business case), the project *must* be declared a success. Does this then mean that the

Table 3.9 The anticipated and actual worth of the new digital pressure controller project

Variable	Parameter				
	Description	Anticipated (as per the business case) "A"		Actual (as per management reports) "B"	
		Magnitude	Timing	Magnitude	Timing
Target outcomes	Gross profit from sale of Y	$10M	Jan-12	$11M	Mar-12
Undesirable outcomes	Cannibalisation of product "X"	$2M	Feb-12	$2.5M	Apr-12
Costs	Total project costs	$4M	Jul-11	$5.5M	Jul-11
Project Worth		$4M		$3M	

project must be declared a failure because the business case was not fulfilled? In this (highly contrived) example, a reasonable answer is "Yes"—implying that the project owner would also be judged as having failed.

3.5 Project Investment Success

We now turn to the evaluation of the investment represented by the project, which is carried out for two different reasons:

1. *Measurement.* To measure the overall performance of an organisation's project portfolio. This requires a corresponding ongoing measurement of performance for each project.
2. *Judgement.* To judge the quality of earlier investment decisions.

This section suggests a test for project investment success.

3.5.1 The Regression Test of Project Investment Success

Project investment success is of concern to the funder who will be asking: "Do I regret having taken the decision to fund this project?" This is equivalent to asking "was this investment successful?"

The discussion in Box 3.5 presents a simple use of the regression test to assess the success of the SIM digital pressure controller example used earlier.

The actual values used in evaluation of investment success must cover all those effects attributable to the funding decision, including fortuitous desirable outcomes (and unanticipated undesirable outcomes). A case study of this kind is presented in Box 3.6.

Box 3.5 Concepts: The Regression Test of Project Investment Success

The Correctness of a Decision versus its Appropriateness

The business case regression test used to gauge investment success is based on an assessment of the end result of the original funding decision. While we use the regression test to judge the performance of the investment (and hence the "appropriateness" of the decision to fund), we cannot use it to judge "correctness" of the original decision. We have here two independent concepts surrounding a funding decision: its "correctness" and its "appropriateness".

A decision to fund a project is based on an appraisal of the business case using some decision rules. The decision rules simply show which of three options is consistent with the information provided in the business case: "Fund", "Don't fund" or "Redo the business case". If a decision was consistent with the organisation's decision rules, then it must be judged as "correct" (regardless of how the project subsequently turned out). So, at the end of a project, we could find ourselves in any of the four situations shown in Table 3.10.

Table 3.10 Possible conclusions that might be drawn from assessment of a project

Correctness of the original funding decision	Judgement about the eventual success of the investment represented by the project	
	Investment successful	Investment unsuccessful
Correct decision	"Just rewards"	"Unfortunate result"
Incorrect decision	"Lucky strike"	"Roosting chickens"

Box 3.6 Case Study: The LAF Business Process Improvement Initiative

Fortuitous Outcomes Result in a Project Investment Success—Despite a Project Ownership Failure

A project undertaken by a large loss adjusting firm (identified here as "LAF Inc") illustrates the way that fortuitous desirable outcomes should be handled in tests of project success. Loss adjusting is the profession involved in the assessment of claims on insurance policies. The processes of loss adjusting are labour-intensive, using highly paid specialists. For some time the loss adjusting profession has been coming under pressure from underwriters to reduce fees, and so profitability has been under threat. A few years ago, LAF responded by undertaking a significant project "Phoenix" to reengineer its core business processes with two target outcomes: reduced operating costs (benefiting the firm itself) and reduced claim processing times (benefiting the underwriters, LAF's clients and their clients—the insured). A business case for "Phoenix" was accepted by the CEO, and the project eventually completed (significantly over budget and over time). Furthermore, the levels of reduction in both operating costs and claims processing time fell significantly short of their targets in the business case.

Near the end of the exercise (in an unrelated move) the major insurance companies went out to tender for their loss adjusting services. LAF's bid for this work was successful, largely because of the emergence (at that point) of fully documented, pilot-tested processes as output from project Phoenix.

The fortuitous eventual outcome "Increased revenue" was, of course, not included in the business case that had been accepted by the CEO (because such an outcome was completely unknown to anyone) but was recognised in the final evaluation.

Despite the fact that project Phoenix was declared a project management failure and a project ownership failure, because the CEO would have accepted a business case that included the (fortuitous) increase in revenue, the project was an investment success.

3.5.2   The Rationale for the Regression Test of Project Investment Success

We claim that the business case regression test is an absolute test of investment success, of which all other tests are special cases (including a number that have extremely undesirable characteristics). Clearly, such a claim requires justification.

Figure 3.6 shows a plane defined by the two variables on which a project's attractiveness is based—worth and riskiness. For the moment, we are working with a rather loose (and undefined) term "riskiness". Later we will replace this with the much more formal (defined) concept of "risk exposure". The plane is divided into two regions separated by a curve identified as the "project investment frontier". Those combinations of worth and risk exposure that lie above (and to the left) of this frontier would be acceptable in a business case, while below (and to the right) should be rejected. Some further comments about the diagram:

- The worth axis could conceivably range from positive down through negative values.
- There will be a certain minimum level of worth for a zero risk project below which a project will not be funded (shown as point "N" on the graph). This accounts for initiatives that are so low in value that they do not even merit the time and effort of processing a business case.

Fig. 3.6 The foundation of the business case regression test

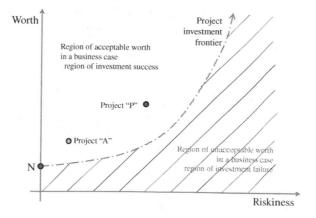

- The project investment frontier will be peculiar to the funding organisation, but, in general, it will take the form of a monotonically increasing curve to reflect the risk avoiding behaviour normally assumed in investment analysis. (The argument presented here does not require this as an assumption).
- While superficially similar to the "Risk-return" functions on which conventional portfolio management theory is based, this chart displays *acceptable* (rather than *available*) investment opportunities for the funding organisation.
- The risk axis can only take on positive values because we have adopted a particular definition of risk that precludes "desirable" forms of risk (refer Sect. 5.4 for a discussion of this concept).

Of course, there are other reasons that a business case might be rejected even if it fell into the upper left region, such as being inconsistent with the organisation's mission statement or behavioural guidelines. In this discussion, we confine our attention to those projects that are acceptable against all other criteria, so that we can explore the relationship between worth and risk.

As well as indicating whether the business case for a proposed project is acceptable, this diagram can also be used to determine (conceptually) whether a completed project was successful. Note that the worth and risk attached to a business case can take on any values, and so the corresponding project could be located anywhere on the figure. For a completed project, the situation is slightly different. Because the bulk of the risk attached to the project will have evaporated at its conclusion, a completed project will normally appear "close-to" the Y-axis, where risk is zero. (It should be noted that we do not require this as an assumption). Residual risk at this point is attached to any remaining uncertainty about benefits being generated into the future from business operations.

Consider the business cases for two almost identical variants of a particular project marked as "P" and "A". "P" is a high worth/high risk version of "A" (which is low worth/low risk). Both are acceptable because they lie above the project investment frontier. Take the following alternative funding scenarios:

1. "P" is available, but "A" is not. Project "P" would be funded and, if it turned out exactly as anticipated, would be declared an unambiguous success.
2. "A" is available, but "P" is not. Project "A" would be funded and, if it turned out exactly as anticipated, would likewise be declared an unambiguous success.
3. Assume that after "P" is funded, it does not go as well as expected, and finishes up instead at point "A" on the chart.

Note that, at "A", the project (although complete) has a non-zero level of risk, implying that a level of uncertainty still remains about the future values of particular variables that were set in the business case. So how is this fact to be reconciled with a declaration that the project is closed (and that target outcomes have been secured)? The declaration of project closure is made with a level of confidence (in the statistical sense) that the target outcomes have been secured.

Given our acceptance of the claim that target outcomes have been secured, how are we to judge project "P"? Clearly, we would have to declare it a success because it is now indistinguishable from "A", which we have already agreed would have been successful.

3.5.3 Qualifying Judgements Based on the Regression Test

When it is directed at investment success, a regression test is triggered by the securing of target outcomes. The conditions under which target outcomes from a project are considered secured require predictions about the future flows of certain variables that then appear in the realised business case. By their very nature, predictions involve uncertainty and so we have to make judgements about success conditional on the accuracy of the predictions involved. In other words, a regression test is, in general, carried out in the face of residual risk. This implies in turn that, in general, the declaration that a project is closed (and that its target outcomes have been secured) must be expressed in terms of *a level of confidence* (in the statistical sense).

The later a regression test is carried out, the larger will be the influence of actual data (and the smaller will be the effects of forecasts). Although this seems to suggest that the test of project investment success should be held off for as long as possible, such a decision would delay the date of project closure and, accordingly is not recommended.

3.6 Comparing the Three Tests of Success

The three success tests (project management, project ownership and project investment) can be summarised in the following way:

- All three use a form of regression testing in which trade-offs are permitted amongst criteria.
- All recognise a criterion called "cost", but each is calculated in a different way. Tests of project management success involve only project costs (they ignore any future operating costs), while investment and ownership success take both components into account.
- All recognise "time", but again, the views are different. Tests of project management success focus on the date by which all outputs were delivered. Project ownership success is concerned with the intended schedules of values for the variables that determine "worth" (as they appear in the business case). Investment success is concerned with the actual schedules of values for the variables that determine worth.

Box 3.7 From the Literature

Project Success in Practice

Measuring project success is challenging, as it means different things to different stakeholders (Lim & Mohamed, 1999). Although few, if any, studies into rates of project success use the regression test, the majority have found that most projects fail (e.g. Johnson, Karen, Boucher, & Robinson, 2001; Luna-Reyes, Zhang, Gil-García, & Cresswell, 2005; Pundir et al., 2007). For example, the Channel Tunnel has a rate of return on investment of negative 14% (Flyvbjerg, 2007).

In a study, described in Appendix C, project success was analysed in more than 700 projects across seven industries in different cultures. Because target outcomes were not identified in most of these projects, a surrogate was used—"funder's satisfaction with the project's outcomes". Funder's satisfaction was found to be the most important project success criterion-rated at almost twice the importance of the iron/golden triangle (Dvir & Lechler, 2004; Lipovetsky, Tishler, Dvir, & Shenhar, 1997). Figure 3.7 shows the mean values of funder's satisfaction from projects across seven industries on a scale of 1 (low) to 10 (high level of satisfaction)

This graph suggests significant differences in project success rates across the different industries. As in project management success, construction organisations achieve the highest scores, while the production/maintenance sector was the poorest performer.

- Project investment and project ownership success also include desirable outcomes that are not used in project management success.
- Project management and ownership success also include outputs—their level of quality and the time they have been delivered, criteria not used in project success.
- It should be noted that if the project is judged as successful then so will the investment.

The three tests of success, compared in Table 3.11, are completely separated, nevertheless it would normally be expected that their *results* would be correlated. In other words, a project failing a test of project management success is more likely to fail a (later) test of ownership success. Because it is difficult to obtain any information about the detail of approved business cases, it is also very difficult to make judgements about project ownership success for well-known initiatives.

Fig. 3.7 Project ownership success levels in different industries

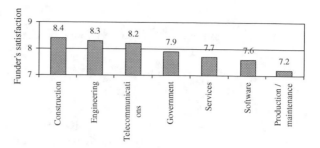

Table 3.11 A comparison of measures used in tests of success

Success dimension	Used in test of project management success?	Used in test of project ownership success?	Used in test of project investment success?
Output scope/quality	Yes	Yes	No
Time	Yes: related to the delivery of all committed outputs	Yes: related to the schedule of all periodic calculations of intended worth	Yes: related to a schedule of all periodic calculations of actual worth
Cost	Yes: related to the outlays incurred in producing, delivering and implementing agreed outputs	Yes: related to the outlays incurred in producing, delivering and implementing agreed outputs and securing target outcomes	Yes: related to all outlays arising from both the project environment and from any associated subsequent operational environment
Undesirable outcomes	Yes: related only to those that can be reasonably attributed to the project manager	Yes: related only to those that can be reasonably attributed to the project owner	Yes: related to all (regardless of cause)
Desirable outcomes	No	Yes: the realised values of all target outcomes	Yes: the realised values of all outcomes (target and fortuitous)

The "Cross-city tunnel" case study (in Box 3.8) confirms that tests of *both project ownership and project management success* are insufficient as tests of *Project investment success*, because they ignore the organisation's investment frontier.

3.6.1 Comparing Approaches to Judging Success

Measuring investment performance is important to a funder (and other key stakeholders) who would like to know whether the investment in the project was appropriate.

Box 3.8 Case Study: The Sydney Cross-City Tunnel

Application of the Three Tests of Success

Section 2.3.3 in Chap. 2 used the example of the cross-city tunnel, built to relieve congestion, pollution and noise in the CBD of Sydney, NSW Australia. Publically available information seems to confirm that the facility was delivered as specified, ahead of time and under budget. Furthermore, assume that there were no detrimental (unexpected, unacceptable, undesirable and avoidable) outcomes. (Note that this assumption does not imply that there were no undesirable outcomes at all).

A target outcome was set as 90,000 vehicles using the tunnel each day. The assumption underlying this outcome is that the bulk of tunnel users would otherwise have made the same trip by driving on surface streets. Operational analysis, however, revealed that fewer than 30,000 vehicles per day have been taken off surface streets. Furthermore, the tunnel had to be sold because the operators became financially insolvent.

One could confidently judge the exercise as a project management success because it met all four performance criteria laid down in the "steel tetrahedron"—scope/quality, timeframe, cost and undesirable outcomes.

It would appear appropriate, however, to judge the investment itself as a failure because we can reasonably infer from all that has happened since that the project would not have been approved by the state government if such an outcome had been known at the time of the original funding decision. Accordingly, the project would be declared an investment failure.

If a project is judged as an investment failure, then, by implication, it must be a project ownership failure and the project owner would be regarded as also failing. It is interesting to postulate a scenario where the project turned out an investment success, but still an ownership failure. This would have been the case if, for example, 65,000 vehicles per day used the tunnel and in Fig. 3.5, the upper curve had an intercept of 75,000 and the lower curve an intercept of 60,000.

While project management performance can be gauged at the time of delivery of the project's outputs, project ownership and project investment success will normally be judged somewhat later. Assume for example, at the end of Project BuyRite, that all outputs have been delivered fit-for-purpose, on time, on budget and without any detrimental outcomes (thus indicating successful project management). The company (ICO) may still have to wait some months before it can determine whether procurement costs have fallen enough to declare the project itself (and the investment it represents) as a success.

Box 3.9 From the Literature

Project Success Measures

As part of an exercise to validate the tests of project success, we have analysed the alignment between the determinants of the proposed project success tests (project management, project ownership and project invest-ment) with those appearing recently in the project management literature. For example, Lechler and Dvir (2010) measure project success using the satisfaction of the project funder ("customer" in the original terminology of their paper). This is a clear case of a specific measure that is readily aligned with the general frameworks of investment success (satisfaction with the investment in project), project ownership success (satisfaction with the work done by the project owner) and project management success (satisfaction with the work done by the project manager) proposed here.

Table 3.12 examines this alignment (the ITO-related terminology has been added in brackets to the original success measure for easy comparison).

The analysis in Table 3.12 suggests that the dimensions appearing in recent literature can be viewed as special cases of those proposed here. The number of different dimensions that can all be considered as specific instances of "desirable outcomes", is particularly noteworthy. While the alignment between the two lists is high, the work of the other scholars does not take into account the possibility of a reduction in project success due to undesirable outcomes.

A second comparison is based on success dimensions discussed in Kerzner (2009) is provided as Table 3.13.

Table 3.13 shows that all Kerzner's seven success dimensions are special cases of the success measures described earlier.

From this analysis we conclude that our framework of benefits and dis-benefits comfortably accommodates all the success dimensions proposed by others.

This approach is consistent with recent discussion of project success dimen-sions (Dvir et al. 2007; Kerzner, 2009; Lechler & Dvir, 2010; Müller & Turner, 2007), where there is agreement on the importance of long term organisational benefits achieved due to the project (see Box 3.9). Amongst the success criteria suggested by these writers are included such end effects as the business impact on the organisation, the opening of new business opportunities for the future, the ability to use the customer's name as a reference, benefits to the community, or any other target outcome stated by the funder in the business case. In other words, all other leading project success studies include success measures that are special cases of the generic framework presented in this book.

Table 3.12 Project success dimensions in Lechler and Dvir (2010) mapped to those introduced here

Project success dimensions (Lechler and Dvir 2010)	ITO success test (to which the proposed dimension belongs)
1. The project had come in on schedule	Project management success (timeframes)
2. The project had come in on budget	Project management success (cost)
3. The project met all technical specifications	Project management success (scope)
4. The results of this project represent an improvement in client (funding organisation) performance	Project ownership success/investment success (desirable outcomes)
5. The project is used by its intended clients (customers)	Project ownership success (output utilisation leading to desirable outcomes)
6. Important clients, directly affected by the project, make use of it	Project ownership success (output utilisation leading to desirable outcomes)
7. Clients (customers) using this project will experience more effective decision making and/or improved performance	Project ownership and investment success (desirable outcomes)
8. The project has a positive impact on those who make use of it	Project ownership and investment success (desirable outcomes)
9. The clients (funders) were satisfied with the process by which this project was completed	Project ownership success (desirable outcomes)
10. The clients (funders) are satisfied with the results of the project	Investment success
11. The project was an economic success for the organisation that completed it	Investment success
12. All things considered, the project is a success	Investment success

Table 3.13 Project success dimensions in Kerzner (2009) mapped to those introduced here

Project success dimensions (Kerzner 2009)	ITO success test (to which the proposed dimension belongs)
1. Within the allocated time-period	Project management success (timeframes)
2. Within budgeted cost	Project management success (cost)
3. At the proper performance or specification level	Project management success (scope)
4. With acceptance by the customer (funder)	Project management success (quality)
5. With minimum or mutually agreed upon scope changes	Project management success (scope)
6. Without disturbing the main work flow of the organisation	Project ownership success (undesirable outcome)
7. Without changing the corporate culture	Project ownership success (undesirable outcome)

3.6.2 Comparing the Many Faces of Success

A further observation is appropriate, related to the timing of the judgments about project management and project ownership success. Project management success is judged when information from outputs closeout becomes available (at the end of

Fig. 3.8 The possible
scenarios that emerge from
the three judgements about a
project's success (*s* success,
f failure)

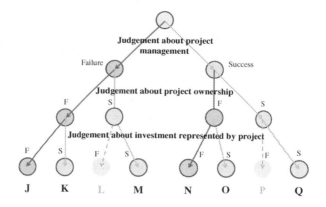

the execution phase). Outputs closeout considers a number of performance measures, including all those that are used to determine the success of project management and many of those used to judge the project owner. Investment and ownership success is judged when information from outcomes closeout becomes available (at the end of the outcome realisation phase). Outcomes closeout considers a (relatively small) number of performance measures, including all those that are used to determine the success of the project as an investment and those needed to finalise a judgement about the project owner. Accordingly, Fig. 3.8 shows examples of well-known projects for four of the six valid combinations of project management, project ownership and investment success.

Some comments on Fig. 3.8 (where the different letters show the scenario to which the comment relates):

L, P   These two cases are invalid (and cannot occur), because when the project is judged as an ownership success, so too must the investment represented by the project (but not vice versa), as suggested by Fig. 3.5.

Q   There is possibly no more memorable or eloquently expressed statement of timeframe for a project than that announced by President Kennedy on 25 May, 1961 before a Joint Session of Congress when proposing (what was to become) the Apollo 11 mission: "First, I believe that this nation should commit itself to achieving the goal, before this decade is out, of landing a man on the moon and returning him safely to the earth." Although available budget information is rather dated and fragmentary, that programme appears to have been not only a project management success, but an outstanding ownership success and investment success.

N   Details of the Cross City Tunnel provided in Box 3.8 make it clear that this project is an example of a project management success and investment failure. It can also be reasonably inferred that it was also a project ownership failure.

K   The Hubble Space Telescope was produced, delivered and implemented with a faulty mirror. That, together with reported budget and timeframe overruns, suggest that the original venture was a project management (and

probably a project ownership) failure. By way of contrast, the general view
of the astronomy community (including NASA itself) is that the images
generated by the instrument have been of inestimable scientific value, and
thus the investment represented by the project was a success (Dunar &
Waring, 1999).

J   A project to develop the (US) Federal Aviation Authority's Advanced
Automation System (AAS) was started in the early 80's. AAS was intended
to consolidate a large number of disparate items of air traffic control
infrastructure. The venture was terminated in the mid 90s having exceeded
its budget and timeframes significantly without delivering core items from
the original scope. This prevented realisation of any benefits and so the
exercise was a failure in both project ownership and project management
terms.

3.7 Critical Success Processes (CSP)

The previous sections introduced the way we gauge success. Achieving success
requires support in the tools adopted for every project. This section introduces a
toolset entitled "Critical Success Processes", an approach that seeks to identify
those processes most important to success across a variety of project contexts.

3.7.1 The Need for an Alternative Critical Approach

The concept of critical success factors (CSFs) has made an enormous contribution
to the project management area. Project managers focus widely and effectively on
the most important project management areas to improve project success. The
identification of CSFs in the project management literature has been a significant
step towards recognising the importance of core project management areas, such
as project planning, top management support and customer involvement. As a
result, a number of effective tools, techniques and models have been developed in
recent years to support these critical areas.

Recently, some criticism has been directed at the CSF approach (Fortune &
White, 2006; Pundir et al., 2007; Zwikael & Globerson, 2006). The critiques,
while accepting the importance of CSFs, also highlight their limitations. The major
identified weaknesses include:

1. *CSFs no longer provide additional practical knowledge for project managers.*
   Project management critical success factors have been studied for more than
   40 years (Rubin & Seeling, 1967). The importance of CSFs is acknowledged by
   most project managers, especially those who already focus on critical areas.
   Moreover, most project management tools and software packages also support

Box 3.10 Concepts: Judging Success

The Special Case of a Subcontractor

Judgements about the success of a contract (and hence the contractor) raise an additional layer of complexity that can, nevertheless, be handled within the framework discussed here.

Consider a new carbon dioxide sequestration project at an existing coal-fired power station run by Hunter Generation and Distribution (HGD), which has committed to a target outcome of reduced CO_2 emissions. Outputs include the new plant, a new operational organisation and new processes. A firm (Carbon Capture and Containment Engineering—C3E) has been engaged as prime contractor/project manager to undertake the entire project for a fixed fee. (HGD is the "principal" and C3E is the "contractor"). How would the eventual success of the exercise be decided?

The concepts introduced in this chapter reveal that deciding "the eventual success of the exercise" involves six separate, independent judgments:

1. The principal's view of the investment represented by the project (of which the contract is a part).
2. The principal's view of the overall project (of which the contract is also a part).
3. The principal's view of the contractor.
4. The contractor's view of the investment represented by the contract.
5. The contractor's view of the contract as a project.
6. The contractor's view of his/her project management (of the contract).

Two points about these judgements are worth noting:

The first is that the six are completely general and apply to all situations where a project involves work completed under a contract. The second is that, while in general, the three judgements made by each party will be independent, in practice, contracts of this type normally link the project management performance of contractors to their remuneration. The intent of this approach is to make the principal's view of the contractor and the contractor's view of the contract congruent.

Before discussing the project from C3E's perspective, we need to consider in more detail C3E's original bid for the work. C3E would have established target outcomes for accepting the contract. Depending on the Company's circumstances, target outcomes could have been associated with a surplus from the job, raised credibility in its market or even the weakening of a competitor's position. All this would have been detailed in a business case to C3E management in support of funding the bid. C3E will judge the investment in the contract with its own (internal) regression test. If, for example, it lost money because an anticipated flow of (what are called in the industry) "Project Extras" did not eventuate, it might well declare the project an ownership and investment failure. If, at the same time the project's outputs were delivered to specification, on time, within C3E's budget and without any detrimental outcomes, then the (internal) project management would have to be judged successful.

Box 3.11 From the Literature

Critical Success Factors for Projects

To improve rates of project success, researchers have explored the main determinants of project success. These are usually called *critical success factors* (CSFs).

CSFs are those areas that if addressed satisfactorily, will contribute significantly to organisational success. Daniel (1961) was the first to introduce the concept of CSFs in the general management area. Since then, the use of CSFs has become widespread in many areas, for example, strategic planning, knowledge management and quality management. The first application of CSFs in the project management arena was made by Rubin and Seeling (1967), who investigated the impact of various factors on project management success. They found that the size of previous projects to which project managers had been exposed was significantly correlated to success. The most cited CSF study has been conducted by Pinto and Slevin (1989). In their research, 418 project managers were requested to evaluate the importance of different factors relating to project success. The study identified the following ten CSFs:

1. *Project mission.* Initial clarity of goals and general directions.
2. *Top management support.* Willingness of top management to provide the necessary resources and authority/power for project success.
3. *Project schedule/plan.* A detailed specification of the individual action steps required for project implementation.
4. *Client consultation.* Communication, consultation and active listening to all impacted parties.
5. *Personnel.* Recruitment, selection and training of the necessary personnel for the project team.
6. *Technical tasks.* Availability of the required technology and expertise to accomplish the specific technical action steps.
7. *Client acceptance.* The act of "selling" the final project to its ultimate intended users.

8. *Monitoring and feedback.* Timely provision of comprehensive control information at each stage in the implementation process.
9. *Communication.* The provision of an appropriate network and necessary data to all key actors in the project implementation.
10. *Trouble-shooting.* Ability to handle unexpected crises and deviations from plan.

Recently, other studies, in different cultures, industries and project typologies, have also developed CSF lists, for example Pinto and Slevin (1988), Morris and Hough (1987), Cleland and King (1983), Johnson et al. (2001), Reel (1999), Soliman et al. (2001). An analysis of these studies reveals that most of these lists share the following factors (Zwikael & Globerson, 2006):

1. *Top management support.* Offered by the senior management of the performing organisation to properly support project management processes.
2. *Project plan.* A baseline document that provides all the information required to make a reliable decision about approving a start to work on a project and track a project's progress.
3. *Personnel recruitment.* Selection of the appropriate personnel for the project team.
4. *Monitoring and feedback.* Collection and analysis of information during project execution used to make decision with the way the project is managed.
5. *Customer involvement.* Active participation of the project owner and selected customers in the project management process.
6. *Project requirement and objectives.* Clear definition of outcomes and outputs.
7. *Adequate spending.* Rigid plan and control of project expenses.
8. *Technical tasks.* Excellency in technical aspects of the development of the new output.
9. *Communication.* The provision of an appropriate network and necessary data to all key actors in the project implementation.
10. *Project strategy.* A high-level plan that focuses on how to achieve project's objectives.

these recognised areas such as project planning and project control. However, CSFs do not advance our understanding of project management success (Soderlund, 2004). They also ignore the history and context of individual projects (Engwall, 2003). Although CSFs may increase the general knowledge of project managers, they are not specific enough to support better decision-making in particular project contexts. For example, the importance of "top management support" is generally accepted, but managers still have little by way of guides and practices regarding how to obtain this type of support.

2. *No one set of CSFs suits all projects.* The contingency theory (Donaldson, 2001) suggests the deployment of choice should be dependent upon variables of firms and environmental conditions. In the project environment, this means that different projects require exclusive managerial focus, for example, while comparing high risk versus low risk projects, project executed in different industries and cultures, different project pace, novelty, complexity and technology (Shenhar & Dvir, 2007). A single list of CSF is unlikely to fit all project scenarios. If CSFs are to be implemented in a particular project, they should recognise its unique features.

3. *Lists of CSFs are not sensitive to different project phases.* The factors that demand special attention from a project manager change throughout the project's life because different project phases have different characteristics. In addition, project managers find that the nature of their work changes from phase to phase. For example planning requires organisation and analysis, while execution demands highly developed people skills. Most CSFs studies present lists of factors that are both static and generic across all project phases. Recent work indicates that unique CSFs apply to each project phase (Lewis et al., 2002). For example, it is generally accepted that a focus on project strategy at the earlier phases of a project is important. Pinto and Slevin (1988) suggest different CSFs that should be applied across the project's life. Table 3.14 analyses the ten most recognised CSFs (Zwikael & Globerson, 2006) in different project phases.

3.7.2 The Critical Success Processes Model

Because traditional CSFs are so general and do not include guidance for project managers' decision making, an alternative approach is required. Some researchers

Table 3.14 The relative importance of critical success factors in various project phases

Critical Success Factors	Initiation	Planning	Execution	Outcome realisation
Top management support	**	**	***	***
Project plan	*	***	*	
Personnel recruitment	**	**	***	
Monitoring and feedback			***	
Customer involvement	***	***	***	***
Project requirement and objectives	***	***	**	*
Adequate spending			***	
Technical tasks	*	*	***	**
Communication	**	**	***	**
Project strategy	***	**	*	

*** Extreme importance
** High importance
* Medium importance

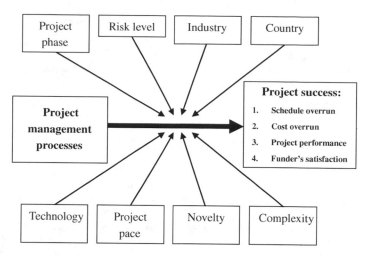

Fig. 3.9 The critical success processes (CSPs) research model

Box 3.12 A Field Study: Critical Success Processes across Project Contexts

The Importance of Different Top Management Support Processes for Projects Executed in Various Industries

We use the CSP model and results, presented in Appendix C, to identify CSPs that appear to be particularly relevant in different project scenarios and phases. Later chapters describe the most critical processes that can be used for each of four project phases: initiation, planning, execution and outcome realisation.

For example, a study based on the model describes in Appendix C (Zwikael, 2008) shows the importance of different top management support processes for projects executed in various industries. Table 3.15 presents the most critical top management support processes in different industries.

According to Table 3.15, the most significant impact on project success in each industry arises from different top management support processes. The results of this study confirm that exclusive CSPs exist in different industries. As a result, one should tailor the type of project involvement to the industry involved in the project, instead of implementing generic best practices. These findings suggest that different project management processes are valued differently by different industry groups. This work indicates that different industries value top management support processes in various project settings.

The full description of the model and the comprehensive results of this study are presented in Appendix C.

Table 3.15 Critical success top management support processes in different industries

Top management support process	Industry
Project-based organisation	N/A
Existence of project procedures	Construction, government
Appropriate project manager assignment	Software, engineering, government
Refreshing project procedures	Software, construction
Involvement of the project manager during initiation stage	N/A
Communication between the project manager and the organisation	Software, production
Existence of project success measures	Software
Supportive project organisational structure	Government, construction, software
Existence of interactive inter-departmental project groups	Software, construction, communications, government
Organisational project resource planning	Software, engineering, government
Organisational project risk management	N/A
Organisational project quality management	Construction
On-going project management training programs	Construction, government
Project management office involvement	Services, software
Use of standard project management software	Software
Use of organisational projects data warehouse	Production
Use of new project tools and techniques	N/A

now accept the need for a more detailed list of critical project management processes. For example, Shenhar, Tishler, Dvir, Liporetstey and Lechler (2002) have identified a number of processes that significantly impact project success, particularly those concerned with developing: a work breakdown structure, a PERT model, a project plan and a quality plan. Raz et al. (2002) have also found that the process of developing a risk plan significantly reduces cost overrun at the end of the project.

While the critical success processes (CSP) model relies on the theoretical foundation of the CSF concept, at the same time it represents a more detailed approach. Instead of providing managers a general list of critical factors, it identifies specific project management processes that most effectively contribute to success in different project scenarios. Executives can ensure that these critical processes are supported by appropriate procedures, templates and tools in the organisation. Project managers can then enhance their effectiveness by devoting more attention, effort and time to these critical processes.

The concept of CSP is more focused, exact and practical than the traditional CSF approach because it recognises the peculiar requirements of different project scenarios. We use CSPs in our discussion of different project scenarios. For this purpose a research model has been developed in which the dependent variables are various forms of project success and the independent variables take the form of project management processes. This list of processes has been drawn from literature covering fields as diverse as project management, organisational support, organisational planning, organisational control and maturity modelling. For more information about this model, see Zwikael and Globerson (2006).

Finally, there is growing recognition that different types of projects require different approaches to their management (Müller & Turner, 2007; Shenhar & Dvir, 2007). Since CSPs are expected to vary across project scenarios, some potential moderating variables are also included in the model. This means that the impact of various project management processes on project success can vary across project scenarios. This also raises the need to identify particular CSPs for different project scenarios, as suggested in the following model.

While it is clear that project management processes impact project success, this relationship may vary among project scenarios. Some variables included in Fig. 3.9 may suggest that various project management processes have different affect on project success across projects. These moderating variables are level of project risk, industry, culture, project pace, novelty, complexity and technology (Dvir, Sadeh, & Pines, 2006; Zwikael, 2008).

Some initial results of this approach have recently been published in the literature. While the identification of project activities was found to be the most critical project management planning process (Zwikael & Globerson, 2006), it has also been found that the level of contribution of different processes to project success varies across sectors. Quality management is the most important process in the service sector (Zwikael & Globerson, 2007), scoping in the information technology sector (Zwikael, 2008) and cost management in the construction industry (Zwikael, 2009).

3.8 Summary

In this chapter, we have assembled a general framework for handing the concept of success as it relates to projects. This involves distinguishing project management success, project ownership success and investments success. In light of this framework we have examined the conventional approach to gauging success in projects and found it wanting, not only as a test of project ownership success and investment success, but also (and perhaps surprisingly) as a test of project management success. By exposing the underlying weaknesses in the accepted approach, we have assembled tests that derive directly from the underlying theory of the ITO model. In the next chapter, we examine the anatomy of the environment within which projects are undertaken. This environment is structured in such a way that projects have the best chances of succeeding at all three levels.

Chapter 4
The Project Environment

4.1 Project Phases

The ITO diagram suggests that projects go through four phases:

1. *Initiation*. Which leads to acceptance (or rejection) of a project proposal. The primary output from initiation is a business case which specifies the project and on which a funding decision is based.
2. *Planning*. Whereby the intended course of the project is mapped out in detail. The primary output from this phase is a project plan on which approval is given to begin substantive work. It is during the planning phase that a suitable project environment is also established. Although the project owner is accountable to the funder for the entire project, the actual planning (and subsequent) delivery of outputs is delegated to, and carried out by, the project manager.
3. *Execution*. When the project's outputs are produced, delivered and implemented. All projects are completely dominated by this phase because this is where most of the resources allocated to the project are deployed. Accordingly, execution takes most of the elapsed time, consumes most of the agreed budget and gets most attention from most stakeholders. All this work has to be managed and administered through a corresponding framework of above-the-line processes. During execution the project manager is accountable to the project owner for delivering all project outputs, fit-for-purpose, according to an agreed schedule, within the agreed budget and without causing any detrimental outcomes (undesirable outcomes that are also unexpected, unacceptable and avoidable).
4. *Outcome realisation*. Where, (through a programme of appropriate intervention), attempts are made to secure the flow of target outcomes. During this phase the project owner is accountable to the funder for achieving such a result. The skills required to guide and facilitate outcome realisation are similar to those that are necessary for the effective management of execution. In most cases, the person who filled the role of project manager will be retained to administer outcome realisation, but will not be accountable for this process.

O. Zwikael and J. Smyrk, *Project Management for the Creation of Organisational Value*, DOI: 10.1007/978-1-84996-516-3_4,
© Springer-Verlag London Limited 2011

Table 4.1 A generic global phase/process structure for projects

The global phase structure of a project
1. Initiation
    1.1. Conceptualisation
    1.2. Business case development
2. Planning
    2.1. Project plan preparation
    2.2. Business case modification (if necessary)
    2.3. Project plan and modified business case approval
    2.4. Project environment set-up
3. Execution
    3.1. Baseline documents maintenance
    3.2. Project monitoring/tracking
    3.3. Project management and administration
    3.4. Project environment maintenance
    3.5. Outputs closeout
This is an above-the-line view of execution, in which the (significant) work surrounding production, delivery and implementation of outputs is not shown explicitly
4. Outcome realisation
    4.1. Facilitation
    4.2. Outcomes closeout

Inevitably, that person will also continue to be identified as the "project manager", even though, strictly speaking the role of project manager concludes when output delivery is complete.

In contrast to the conventional view (e.g. PMI 2008; OGS 2007; Kerzner 2009), we propose that the span of a project's life does not end with delivery of outputs (at the end of execution), but that it continues until target outcomes are secured. Given the importance of outcome realisation in projects, a fourth project phase has been introduced for that purpose. Another difference with existing models is our treatment of closeout: the ITO model implies that closeout is required after both the delivery of outputs and securing of outcomes. As a result, closeout processes are included in two of the project phases.

Table 4.1 shows a proposed generic structure of global phases for a project. This represents a high-level description of the work that is common to all projects. In this figure, each of the four project phases, introduced above, is divided into more detailed processes.

As the hierarchy in Table 4.1 suggests, these global phases are broken out into a number of processes that are also common to all projects.

4.1.1 Project Initiation

Initiation is, in effect, a project specification phase which has two distinct processes: "conceptualisation" (where the foundations of the project are established) and "definition" (where its key parameters are set).

Conceptualisation starts when someone triggers the idea for a project. If the idea is accepted in principle, then formal specification of the initiative follows with development of a business case. A business case provides only enough information for a funding decision, accordingly it includes a comprehensive discussion of outcomes, outputs, costs and risks. The document does not normally contain such detail as the proposed communications plan or who will be doing specific tasks on particular dates—that sort of information is addressed in the project plan.

At its inception, each project has a "champion"—someone who wants to see the project funded and who will also lead the exercise through to the point where a business case is accepted by the funder. A champion is recognised in either of two situations: as the person charged by the funder with responsibility for assembling the business case or as the person who gains approval from the funder to assemble the business case. While the champion drives initiation, it is the funder who will make the final decision to accept or reject the business case. In most cases, the champion will not have the capacity or the capability to assemble the business case him/herself, and so the work is normally assigned to someone who is appropriately skilled. That person is clearly a candidate to fill the role of project manager, and might well be described as the "project manager designate". In that event, it should be noted that the business case is still owned by the champion, who will eventually present it to the funder. In assembling the document, the project manager designate is simply supporting the champion and at no time becomes accountable for its tabling or acceptance.

When the funder accepts a business case, he/she will normally commission someone as his/her agent to assume responsibility for the project overall. That person becomes the project's owner, accountable to the funder for realising the business case and hence eventually securing the flow of target outcomes. There is nothing to prevent the funder from acting as owner. In most situations, however, funders are responsible for such a large portfolio of projects that it is not feasible for them to give each one the close attention it demands, and so it becomes necessary to delegate them to other senior managers. See Chap. 5 for a detailed discussion on initiation. At this point, the project owner will usually initiate the engagement of a project manager.

4.1.2 Project Planning

During this phase, the project manager develops the project plan, which includes, amongst other things, a strategy for ensuring that outputs are fit-for-purpose, a comprehensive model of all the work that is anticipated, a schedule of milestones by which critical tasks will be completed, information on the resources and funds required, and a description of the way communications are to be undertaken during the project. This significant piece of work demands appropriate resources and time. The construction and engineering industries, in general, recognise the need for high quality, complete and robust project plans, which may go some way towards explaining the relatively high levels of success they enjoy (see some related study

results in Appendix C). Unfortunately, many business projects are started with
unreliable and incomplete plans (or worse still, with no plans at all!). During the
development of the project plan, project managers should also identify the critical
processes on which they are going to focus during project execution. For example,
while in one project a focus on quality assurance may contribute to the best results,
in another, strict schedule control may be key. The selection of critical success
processes for a project is dependent upon the type of culture, industry, organisation
and stakeholders involved. Once the project plan has been approved, it is important
that this be acknowledged and accepted by the project owner and members of the
steering committee (acting on behalf of the funder).

Before a start can be made on the project's outputs, some preparation is
required, in the form of a set-up activity. Here the project environment is established by marshalling the team, implementing the governance model and, in some
cases, by assembling the project's own infrastructure (such as offices, vehicles and
systems). Once this has been done, then execution proper can begin. See Chap. 6
for a detailed discussion on planning.

4.1.3 Project Execution

During this phase, the project's outputs are created, by undertaking the work
described during planning. The processes of execution are guided by somewhat
tentative models that are usually assembled with incomplete information. (Because
projects are not repeated, everyone who is involved faces uncharted territory). It has
to be expected therefore, that these models will require continuous ad hoc modification and refinement as the work unfolds. Control and management of that type of
change is based on a separate (but closely connected) stream of above-the-line
activity, mainly concerned with monitoring, tracking, reporting and guiding.
Following delivery and implementation of the project's outputs, a closeout workshop is normally conducted to expose opportunities for future improvement of the
organisation's project performance. If outputs are delivered progressively over an
extended period of time (as in the case of a large project), then a succession of
closeout workshops may be appropriate. Closeout is strictly above-the-line, concerned with judgements about the planning and management of the project. Closeout
is quite distinct from a conventional post implementation review (PIR) which has a
below-the-line focus, concerned primarily with judgements about the operation and
quality of project deliverables. See Chap. 7 for a detailed discussion of execution.

4.1.4 Outcome Realisation

This phase represents a significant extension to the work that has traditionally
defined "the project". In terms of the ITO model, target outcomes are generated by
utilisation of outputs. Outcome realisation (for which the project owner is

accountable) is that portion of output utilisation that occurs up until the flow of target outcomes has been secured. Beyond that point, the generation of target outcomes takes place as part of regular business operations (for which a line manager is accountable). Outcome realisation ends (and the project concludes) when, within an agreed timeframe, flows of desirable outcomes stabilise below, at or above the target thresholds that have been set. Outcome realisation involves two major processes:

1. *Facilitation* where a support group will attempt to influence the generation of target outcomes with an appropriate programme of intervention. Facilitation ends with release of some form of "end-of-project declaration" confirming not only that target outcomes are secured, but also that they likely to continue flowing into the future.
2. *Outcomes closeout* where key stakeholders review outcomes achievement and make a judgement about the success of the project. Output from this process is an outcomes closure report that analyses the benefits gained from the project.

It is important to note that the project manager and team discharge their responsibilities when outputs are implemented. Some members of the team (and even the person who acted as project manager) may remain during outcome realisation, but their role is now quite different, simply supporting the project owner and customers. See Chap. 8 for a detailed discussion on outcome realisation.

4.1.5 Accountabilities During a Project's Life

The foregoing discussion suggests that primary project accountabilities shift from phase to phase, as is described in Fig. 4.1.

Figure 4.1 shows that the project manager is only one of several stakeholders who hold accountability in different phases. In contrast to what sometimes happens, a project is not a one-person show (featuring the project manager), but a collaborative effort involving many key players: the project champion (for project initiation and the development of the business case), project manager (for planning, execution and output delivery), project owner (for outcome realisation and a secured flow of outcomes in the future) and line manager (for any ongoing generation of outcomes).

The diagram also shows how, in some cases, the four phases of a project eventually link into routine business operations. This happens when outcome generation is intended to continue into the future.

4.2 The Anatomy of a Project

In addition to progressing through phases (a time-based perspective), a project can be shown as made up of elements representing a structural view.

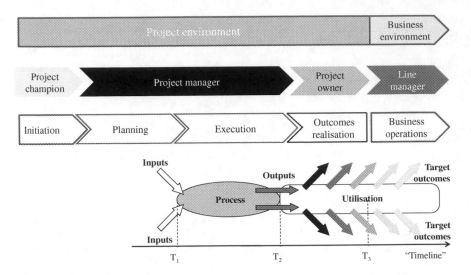

Fig. 4.1 Direct accountabilities during a project's life

4.2.1 The Elements of a Project

All project methodologies employ a taxonomy of core elements which frame their management guidelines. For example, PRINCE2 (OGC, 2007) recognises: organisation, plans, controls, business case, risk, quality, configuration management and change control. The PMBOK (PMI, 2008) identifies nine knowledge areas: integration, scope, time, cost, quality, human resources, communications, risk and procurement. If it is to be effective, at its highest level, a taxonomy of elements should be exhaustive and exclusive while allowing for the addition of other levels that eventually provide a "slot" for every concept used in the associated discourse. Unlike other methodologies, where the classification of project elements appears to arise from a largely ad hoc grouping of "accepted practices" (a bottom-up approach), we suggest an alternative top-down approach using a taxonomy of project concepts based on seven elements. These elements, in turn, are drawn from various components of the ITO model, as shown in Table 4.2.

1. *Scope*. In order to generate target outcomes, particular outputs have to be delivered with specific features and then utilised. Scope is covered in Sect. 5.2 in Chap. 5 (because it represents the foundation of a new project, which is the topic of Chapt. 5).
   A project's success is sensitive to the outputs that define its scope, therefore scope must be set, specified and validated and, accordingly, is included as a structural element.
   All project outputs should have features that make them "utilisable" by project customers. When an output has such features it is said to be fit for purpose. Unless outputs are fit-for-purpose, they cannot be utilised effectively by a project

Table 4.2 The linkage between the project elements and components of the ITO model

Project element	Component of the ITO model				
	Inputs	Process	Outputs	Utilisation	Outcomes
Scope		Outputs produced	Outputs listed	Fitness for purpose features inferred	
Stakeholders		Roles established		Project customers identified	Beneficiaries identified
Governance		Organisation defined			
Schedules			Schedule for outputs developed	Schedule for output utilised	Schedule for outcomes generated
Resources	Resources required for the project	Resources utilised			
Risks		Risks mitigated			
Issues		Issues resolved			

customer (and so cannot contribute to the generation of target outcomes). We are thus led to an important principle of project management: a project is scoped if and only if its outputs are defined. A project's outputs are defined when two conditions are met:

- The outputs are all listed.
- Each is supported with a list of critical fitness-for-purpose features.
  Scoping is central to project management because it rigorously establishes a reliable list of outputs, without which target outcomes cannot be generated. Once outputs are identified, it then becomes possible not only to plan the project (and in particular to describe the work required for the production of these outputs) but also to manage any changes in scope during project execution.

2. *Stakeholders*. Various entities can influence the way the project unfolds. Stakeholders are discussed below (Sect. 4.5) because they have such an important impact on the project environment (also covered in this Chapter). Stakeholders, by definition, can influence a project's success. Therefore the nature of this influence should be analysed and managed, and so "stakeholders" are included as a structural element.
For example: the funder identifies (or commits-to) target outcomes, the project team (including external suppliers) delivers outputs and customers utilise these

outputs to generate target outcomes. It is important, therefore, to make sure all these entities (and a number of others as well) are engaged appropriately, and that none of them is able to hinder achievement of the funder's objectives. Managing these stakeholders and their expectations is critical to project success.

3. *Governance*. A project is a form of investment and, accordingly, must be managed if it is to be successful. Effective management demands the assignment of roles, responsibilities and accountabilities. The standing organisational arrangements in most organisations do not accommodate projects very well, therefore special organisational models are required, and so "governance" is included as a structural element. Governance is covered in this Chapter (Sect. 4.3) (because it relates to the organisation of the project environment).

By definition, projects are not repeated and so they represent a temporary structure within the organisation which is disbanded when outcomes are secured.

4. *Schedules*. As will become clear in Chap. 6, the work associated with a project has to be analysed so that reliable dates can be attached to critical points in the project's life, particularly those relating to delivery of outputs and realisation of outcomes.

A project is a process which requires a "script" if it is to be executed successfully. An effective script for a project can take any of a number of forms, the most common of which involve a WBS and Gantt chart (together identified as a "schedule"). The need for an executable model of the work in a project is reflected in our recognition of schedules as a structural element. Scheduling is covered in Chap. 6 (because it represents one of the most significant aspects of planning—the core of Chap. 6).

5. *Resources*. Outputs require work and work demands resources, for which we use a simple two-way classification of labour and money. Resourcing is also covered in Chap. 6 (because it is also a significant aspect of planning).

Cost is one of the three variables in a project's "equation of worth". Costs are driven by a number of factors, including the resources required to undertake a project's work. These resources (inputs) should be managed throughout the project's life, from initiation through outcome generation, and so are accepted a structural element.

6. *Risks*. Projects are subject to all sorts of uncertainty which must be managed. Project uncertainty is formalised as risk. Understanding the level of risk to which a project is exposed is a precondition for accepting the business case. Risk is covered in Chaps. 5, 6 and 7 with discussions about gauging project risk, risk mitigation planning and risk control respectively. This approach that we have taken accepts that risk management is best integrated throughout the project management process, rather than isolated as a separate process (as some say "project management is all about reducing the level of risk").

A project portfolio can be viewed as a list of investment opportunities, ranked according to their attractiveness. A project's attractiveness is determined by its worth and riskiness. Riskiness is a gauge of a project's exposure to falling short of its target worth. A framework for the management of risk is, therefore included as a structural element.

The risk element of a project provides the frameworks for identifying factors that threaten achievement of the business case, analysing them and selectively mitigating their effects.

7. *Issues*. The project environment is also subject to other sorts of developments that, while not representing a threat to success, demand a response. We label these events as "issues" (which should not be confused with "realised risks"). Issues are covered in Chap. 7 because they represent an important part of the ongoing management of project execution—the topic of Chap. 7.

Projects face a continuous barrage of spontaneous events, most of which are low-scale in nature. In some cases these events, and the reactions to them by key players, result in a progressive evolution of the project environment. A framework for the management of issues, therefore merits inclusion as a structural element.

4.2.2 The Types of Work in a Project

Work undertaken in a project is of two distinct kinds. The first (and most obvious) is required for the production, delivery and implementation of those outputs that are to be utilised by a project's customers (eventually leading to the generation of target outcomes). We call this below the line work. The artefacts emerging from this work are accordingly identified as below-the-line outputs.

In addition, there is the work of planning and managing the project's below-the-line work. We call this above-the-line work. Above-the-line work gives rise to above-the-line outputs, such as schedules of milestones, project progress reports and the risk register. Above-the-line work is associated with the seven structural elements discussed above.

Figure 4.2 shows how the two classes of work are related. Below the line work is concerned with the project's deliverables. Above the line work is concerned

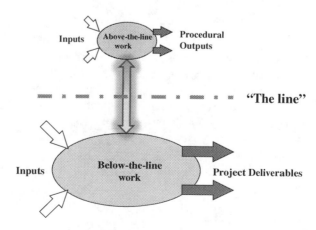

Fig. 4.2 The relationship between "above-the-line" and "below-the-line" work

with planning and managing (controlling) the project's below-the-line work (Boxes 4.1, 4.2).

Box 4.1 Terminology in Other Disciplines

Above and Below the Line

The terms "above-the-line" and below-the-line" are used (with unrelated meanings) in marketing, accounting and contract management.

In the marketing profession, the terms are used to classify sales promotion activity. "Above-the-line" refers typically to mass media advertising. "Below-the-line" activity is transaction-oriented (such as price discounting, coupons or even point-of-sale promotion). The distinction is important to advertising agencies where below-the-line promotions (under normal conditions) earn no commission, while above-the-line activity does.

The accounting profession uses the terms to distinguish transactions that impact an organisation's income statement ("below-the-line") from those that affect its balance sheet ("above-the-line").

The contract management profession uses the terms to classify the costs incurred in contracts. There, "below-the-line" refers to those factors that can impact the value of a contract after the basis of an estimate has been fixed (and agreed). Below-the-line items include those factors that lie beyond a contractor control, such as escalation and exchange rate variation. By way of contrast, above-the-line refers to factors that, under normal conditions, have no implications for the value of the contract.

4.2.3 A Project's Baseline Documents

'Above the line' work is associated with the development of the many documents that are necessary for the management of a project. Throughout its life, the shape and form of a project are governed by its baseline documents. A project has two baseline documents: a business case and a project plan which together establish the project's objective, scope, constraints and ground rules. In addition, baseline documents guide all significant aspects of a project's strategy, processes and decisions, as well as setting out the roles of all key players. Nothing of significance should be effected unless it is consistent with these "frames of reference".

The business case and project plan are discussed comprehensively in Chaps. 5 and 6 respectively.

> ## Box 4.2 A Practicality: Estimating Labour for a Project
>
> ### *How Much Project Load Arises from Above-the-Line Work?*
>
> A common rule of thumb (for which there is strong supporting anecdotal evidence) is that the ratio of internal labour consumed by above the line work to the total internal labour consumed by the whole project is about 15%. (Note that this ratio involves only the labour deployed or hired under contract by the organisation undertaking the work, it excludes the cost of purchased products).
>
> If, for example, Project BuyRite is expected to consume about 15,000 person-hours of ICO staff and contractor time, then another 2,650 person-hours (or so) will be consumed in all above-the-line work across the life of the initiative.
>
> This now raises a question "What happens to this ratio as the project size increases?". *A priori*, a plausible argument can be advanced suggesting that diseconomies of complexity grow faster than economies of scale. If that conjecture is right, then the ratio of above the line effort to total project effort would rise with increasing project size.

4.3 Project Governance

A project usually involves a significant amount of work and a large number of people, hence some sort of management and administrative framework is required to organise all of those who are involved. Such a framework is called a project governance model. An organisation's standing management arrangements are normally inadequate for project governance because they are concerned with accountabilities and reporting lines that rarely bear any relationship to those required by a project.

4.3.1 Overview of Project Governance

A project governance model has two parts. The first is a diagram identifying all of the entities that will be recognised in the project, and showing how those entities relate to each other. The second is a set of supporting descriptions of the roles to be filled by each entity.

The range of feasible management models that could be proposed for a project is very wide indeed. Casual observation of the way that typical business projects are organised suggests that many of the organisational structures adopted in practice are largely ad hoc with poorly defined roles leading to gaps/overlaps of responsibility, conflict amongst participants, inadequate processes and unreliable decisions. A governance model should be not only theoretically sound, but also have predictable behaviours to support project success. It is appropriate therefore to constrain the optional governance arrangements that might be considered for a project. The principles discussed below constrain the generic governance model for projects in three ways: by setting its general shape and features, by identifying the particular elements that are used in the proposed templates for the role definitions and finally by guiding the way roles are defined for specific entities.

4.3.2 Principles of Project Governance

The approach to project governance described here is based on principles which are discussed at two levels: those on which the entire model is based and those apply to particular parts of the model. In the following discussion these principles take the forms of a *general* list (related to the overall model) and *specific* lists (related to particular roles) respectively. It should be noted that if we locate projects on some spectrum of size/complexity, as we move to the small end of the scale the need for a formal project governance model fades (and the need for these principles also evaporates).

1. *Avoidance of conflict of interest.* When assigning project roles, conflicts of interest must be avoided.

   - Description: A conflict of interest arises when someone could benefit (even indirectly) by acting against an assigned responsibility. It should be noted that acting improperly and facing a conflict of interest are governed by separate principles (someone can face a conflict of interest without acting improperly).
   - Example: An external supplier is disqualified from participation in many of the decisions faced by a steering committee (such as scope change) and therefore cannot be a member of that entity. (This does not, of course, prevent a supplier from joining other entities such as a reference group or project control committee).
   - Foundation: Avoiding conflict of interest is a common general principle adopted in many areas of commerce and law (e.g. Simon, 1995).
   - Rationale: Someone facing a conflict of interest may make decisions that do not favour the project. The principle reduces the risks associated with such an event.

2. *Projects are recognised only implicitly in an organisation's "standing structure".* Project governance models are "hooked into" an organisation chart through the operational roles of steering committee and reference group members. Structure charts for each project should show the world from the

project's perspective, rather than from the point-of-view of the funding or performing organisation.

- Description: A project governance model "reports in" to an organisation via those assigned management and staff who retain roles in the organisation's "standing structure". It is unnecessary (and undesirable) to show project governance models as extensions to a standing structure chart. Members of the steering committee in particular provide the organisational connections required by a project governance model.
- Foundation: Project-related accountabilities and responsibilities lie with those stakeholders who have been commissioned to fill specific roles in the project governance model. All these roles eventually report into the steering committee whose members are appointed by the funding organisation(s). Reporting arrangements into those organisations will be established as part of their appointment. In some cases, similar reporting arrangements will also be put in place for certain reference groups and advisers.
- Rationale: A project governance model is intended to show how the roles within a project are arranged rather than how the projects within an organisation are arranged.

3. *A project owner is to be held accountable by the funder for target outcomes.*

- Description: project owner(s) are appointed by, and held accountable by, the funder(s) for the eventual realisation of target outcomes. A funder may elect to fill the role of project owner him/herself. There can be multiple project owners. Because multiple project owners implies multiple "clients" for a project manager, is becomes necessary to provide a framework by which different views and directions can be reconciled in cohesive, consistent instructions. The steering committee provides such a framework.
- Example: In Australia, a major programme of projects is being undertaken to reduce the incidence of child abuse. The various participating states and territories have consistent (but distinct) target outcomes, for which each has made someone senior accountable. All project owners are members of a steering committee which agrees on instructions for the project manager.
- Foundation: The principle arises from the ITO model, coupled with the principle of separation of purchaser from provider (following). An implication of the principle is that a project manager cannot be held accountable for target outcomes.
- Rationale: By making someone accountable for outcomes we ensure that the interests of the funder drive the conduct of the project.

4. *In a project, a "purchaser" is to be identified and separated from the role of "provider".*

- Description: in any transaction two separate entities should fill the role of purchaser and provider.

- Example: The project "purchaser" (funder/project owner) should be separate from "provider" (project manager). While these two roles should not be filled by the same person, these people could, and generally will, come from the same organisation.
- Foundation: The principle (known formally as "the separation of purchaser from provider") is used widely, especially in reform of government services for which there is no true market (such as public health). Turner and Muller (2004) use (a related) principal-agent theory (Eisenhardt, 1989) to analyse this relationship in the project environment.
- Rationale: The principle encourages clarity and transparency of the processes surrounding a transaction by separating accountabilities of a "supplier" and a "customer". The relationship between the project manager and the project owner serves as the nucleus for the entire governance framework.

5. *A project manager is to be held accountable by the owner(s) for delivery of project outputs.*

- Description: a project manager is notionally engaged by, and held accountable by, the owner(s) for the production, delivery and implementation of the project's outputs.
- Foundation: The principle arises from the existing practices of the project management profession (e.g. PMI, 2008: p. 369).
- Rationale: By making someone accountable for outputs we can exercise considerable control and influence over the direction of the project.

6. *The role of each project participant is to be formally defined and acknowledged as a condition of appointment.*

- Description: Every significant role recognised in the project governance model is to be specified in some formal instrument (such as a charter, role description or terms of reference) which incumbents are required to acknowledge and accept. There will be a threshold of involvement below which it might be seen as incidental, in which case the principle can be ignored.
- Example: All members of the steering committee would be expected to accept a (fairly generic) charter that covers (amongst other things): the purpose and role of the committee, the scope of involvement in the project, and a confirmation that members will work towards the successful achievement of the business case.
- Foundation: The principle is grounded in organisational theory (Daft, 2007) and human resource management literature (Dressler, 2008).
- Rationale: Formalising roles becomes, in effect, a mechanism for managing the risks associated with organisational overlaps and gaps. It also reduces the risks associated with misunderstanding of roles.

7. *The release of a resource to a project by a third party should be covered by a formal agreement.*

- Description: Where the appointment of a person to fill a project role requires the approval of another entity, then it may be appropriate to have the arrangement formalised with an appropriate instrument. If the approving organisation is a separate legal or commercial entity, then the instrument would take the form of a contract or commercial agreement. If the approving organisation is not a separate legal or commercial entity, then the instrument would take the form of a memorandum of understanding (MoU). An MoU lays out the terms of an agreement, but without the legal sanctions that underpin a commercial contract.

- Examples: In Project BuyRite, Lindsay Thomas (ICO's experienced Purchasing Manager) will obviously play a critical role in documenting current procurement process, and in the design of the new versions of those processes. It would be appropriate for Paul Myer (BuyRite project manager) and Nancy Palmer (National Procurement Manager) to draft an MoU governing Lindsay's deployment to the project. In a similar way, the engagement of Process Re-engineering Services Inc (a firm of process improvement specialists) would involve a commercial contract.
- Foundation: The principle is enshrined in long standing business practice and commercial law (e.g. Kerzner, 2009).
- Rationale: Formal agreements ensure that everyone involved in the arrangement is clear on what is being offered, the conditions under which it is offered and what processes are to be followed in the event that conditions of the agreement are violated.

4.3.3 The Structure of the Project Governance Model

Numerous models and structures have been proposed at different times and in different methodologies for the organisation of projects. Here we attempt to provide a completely general governance framework based on the principles defined above and made up of four divisions, as presented in Fig. 4.3: steering, delivery, reference and assurance. Most, if not all, of the approaches in common use can be mapped to this model as special cases.

Fig. 4.3 The four divisions of the generic project governance model

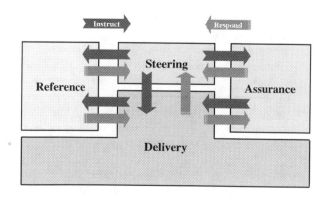

Two divisions (steering and delivery) must appear in the project governance model of every project. The reference and assurance divisions are not always necessary. The project governance models of projects differ therefore, from each other at three levels: in the need for the reference and assurance divisions, in the structure of the reference division (when it exists) and in the membership of the particular entities that are created in each division. It should also be noted that the governance model for a particular initiative will evolve over the life of the project in these same three levels. In other words, changes over time will be noted in the number of entities represented within each division and in the membership of on-going entities. For example, it is conceivable that a reference group may exist only for a single working session.

1. *Steering.* There are two (closely related) entities in this division—the project owner and the steering committee. The steering committee approves significant changes in either of a project's baseline documents, especially parameters that impact worth (such as outcomes, timeframe and cost) and directs the project towards achievement of those baseline parameters. All projects must have a project owner, but not all projects need a steering committee—in that case the project owner becomes a "steering committee" of one.
2. *Delivery.* In this division are located all those responsible for producing the project's outputs and also those responsible for administering delivery. There are three (again, closely related) entities here: the project manager, his/her project team, including (when required) an administrator (or administrative team). The project team may, in turn be made up of many component entities such as sub-teams, contractors, consultants and suppliers. The structure adopted for a team will depend on the peculiarities of the project and could be based on: skills, functional areas, outputs, contracts or type of work. The larger and longer the project the more it is likely to evolve over time. An administrator is only needed when the above-the-line workload exceeds the equivalent capacity of an individual team member. In large projects the administrative load may become so great that it requires its own support team.
3. *Reference.* In this division are located those who have been commissioned to provide specialised input to either the team or the steering committee. There may be many entities in this division. Those that have a number of members will be given titles such as "reference group" or "working party", while those that involve individuals will usually be identified as "advisers".
4. *Assurance.* In this division are located those responsible for independently monitoring the conduct of the project on behalf of the steering committee. The two most common assurance roles are those of project assurance counsellor and probity counsellor. A project assurance counsellor is engaged to ensure that the project is conducted in accordance with the project management framework adopted by the funding and performing organisations. A probity counsellor is appointed to ensure that the commercial dealings between project participants and the outside world are being managed in accordance with the commercial guidelines of the funding organisation. Not all projects require these roles.

The relationships between these divisions bear further discussion because they determine how all entities in a specific instance of a model eventually interact.

The steering committee is accountable only to the funder(s) (a role not recognised explicitly in the project governance model). The funder is the "ultimate" entity in the project governance model. The steering committee approves the detailed design of the project governance model and so all entities in a project governance model are eventually accountable to the steering committee for successfully discharging their responsibilities.

Each entity in the reference division "reports" (in the line sense) to either the steering committee or the project manager. While the project manager, for example, may *brief* a reference group at its regular meetings, he/she is never subordinate to such a group (otherwise such a group would become a "competitor" to the steering committee).

Although assurance counsellors (in those projects where they are appointed) can *instruct* the project manager to provide required management information about the project, in practice they will be working closely and cooperatively with the project manager and so any information for quality assurance purposes is more likely to be *requested*.

The project team (through the project manager) is closely linked to the steering committee by formal reporting arrangements including regular review meetings.

4.3.4   Classes of Entity in the Project Governance Model

Within the four divisions are six classes of entity, as shown in Fig. 4.4. The steering committee and the project owner belong to the "steering" division, the project manager, the project team (including any assigned project administrators) are included in the "delivery" division, and the other two divisions share one class of entity each. All these are closely related to the nine generic classes of commissioned stakeholders discussed below. In other words, anyone who fills a role in the project governance model, does so by appointment or invitation and so becomes a stakeholder in the project because of that engagement. One obvious exception concerns the project champion who is involved before the governance model is designed, but not recognised as a role beyond initiation. In the following discussion, when referring to the six generic classes of entities that lie within the various divisions, for simplicity we call them elements. Particular instances of each element are represented by entities (individuals or collectives).

4.3.4.1   Steering Committee

Role: The steering committee supports the project owner in looking after the interests of the funder. It seeks to ensure that the business case/project plan will be realised and that project is always 'pointed towards' its target outcomes.

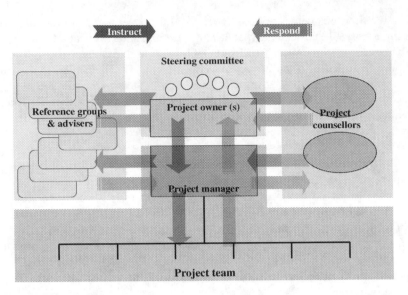

Fig. 4.4 The detailed project governance model showing the entities within each division

Principles.

- There can be only one steering committee, made up of a small group of powerful supporters of the project. An implication of this principle is that anyone who opposes the project is automatically disqualified from membership. (If an opponent of the project is to play a role at all, it would be as a member of a reference group).
- The steering committee works almost exclusively above the line and is not an alternative or (worse still) a competing project team. Consider a project where disagreement has broken out amongst team members over the design of a high-quality (and expensive) brochure. It would be appropriate to identify this as an issue for consideration by the steering committee (from whom guidance might be sought on how the matter is to be resolved), but it would not be appropriate to have the steering committee make the design decision.
- A steering committee is not there to "represent" specific interest groups (internal or external), reference groups are the appropriate forums for dealing with that sort of issue.
- The project owner is a member of the steering committee. Despite this, it will usually be the project owner who tables proposals for changes to baseline documents and seeks the steering committee's approval for those changes. There is an alternative principle under which the project owner is not a member of the steering committee, but instead reports to it. In that case the steering committee becomes solely concerned with the funder's interests. Although a valid model, that variant may weaken the accountability of the project owner for securing target outcomes, and is not discussed further here.
- While it is common for membership of steering committees to change over the life of the project, a succession of short-term appointments is highly undesirable

because it can lead to a steering *committee* degenerating into a steering *parade*. It is reasonable to expect that projects will be no more stable than their steering committees.

Membership.

- A funder can be a member of a steering committee.
- The role of the steering committee is such that no one should be appointed as a member in order to gain access to his/her technical expertise. That sort of involvement would be best accommodated by making the participant an advisor or a member of a reference group (or even a member of the project team).
- A trade-off has to be made with steering committee size. The larger the committee, the slower and more cumbersome the decision cycle. If, on the other hand, key players are missing, then the ability of the committee to positively influence the course of the project may be compromised.
- Because the steering committee will, in consultation with the project owner, approve changes to the project's baseline documents (especially those involving scope, timeframe and cost), membership is restricted by consideration of conflicts of interest. Not only are external suppliers clearly disqualified from membership, but there could well be cases where such a restriction might also apply to internal suppliers as well.

Qualifications. A steering committee member should be strongly committed to the success of the project. In addition he/she should be able to add value to the work of the committee by being politically influential, a reliable decision-maker and/or a strong leader. Membership of a steering committee is onerous, but, in general the role is poorly understood amongst the executive management community. It is highly desirable to have members of steering committees exposed to some form of formal induction programme (even if only short) to ensure that members understand the role and are able to participate in core processes.

Engagement. A charter is established for the steering committee (based on a generic structure that would apply to most projects). The project funder would invite candidates to join (or appoint them). Each candidate is required to formally accept the steering committee charter. Assembly of the steering committee would normally be done during project planning under the leadership of the project owner with administrative support from the project manager.

Performance evaluation. A steering committee should be assessed collectively by the funder(s) after a project's target outcomes are secured. That assessment should be based on the extent to which the last baseline business case was realised.

4.3.4.2 Project Owner

Role. The project owner is fully accountable to the project's funder for realising the business case in general and securing the flow of target outcomes in particular.

The project owner is, effectively, the funder's agent. Ideally the project owner will chair the steering committee, but in some cases there may be better-qualified members to fill that role.

Principles.

- In all but the most extraordinary of circumstances, project owners would normally be appointed from inside the funding organisation.
- There can be more than one project owner, however each must be appointed by a funder. In many such cases, the project owner may be identified as a programme owner.
- The project owner is the project manager's client. In the case of multiple project owners, the steering committee becomes the effective client of the project manager.

Qualifications. A project owner should be strongly committed to the project and have a good understanding of project planning and management. A project owner must be capable of dealing with the numerous political issues that arise in the course of a project.

Engagement. A role definition is established for the project owner (based on a generic structure that would apply to most projects). He/she is engaged by the funder. The project champion is, in many cases, an obvious candidate for this role. The project owner is required to formally accept the engagement. This would normally be done by the time that a decision is taken in favour of the business case.

Performance evaluation. The project owner should be assessed by the funder(s) after a project's target outcomes are secured. That assessment (as in the case of the steering committee) should be based on the extent to which the last baseline business case was realised.

4.3.4.3 Project Manager

Role. The project manager is fully accountable to the project's owners for delivering the project's committed outputs, fit-for-purpose, within the constraints of an agreed budget and timeframe and without causing detrimental outcomes.

In large projects a management forum might be created, made up of team leaders, sub-project managers or representatives from suppliers. Forums of this kind go under many titles including: project management committee, project control group or project board. (The last term would be undesirable in a model based on the framework presented here because it may be seen as having outcomes-related responsibilities). Ideally the project manager will chair such a group, but in some cases there may be better-qualified members to fill that role.

Principles.

- Project managers can come from inside or outside the performing organisation and if from outside may well be appointed on a fixed-term contract.

- The project manager is engaged by the project owner.
- There can be only one project manager, however under him/her there may be multiple sub-project manager or team leaders, each one responsible for particular outputs and/or certain areas of work. In many such cases, the project manager may be identified as a programme manager and those reporting as "project managers". If this principle is violated (and there is more than one project manager) then the question arises who is accountable for resolving conflicts about output delivery. The answer, by definition, identifies a project manager (and so logically there can be only one).
- The project manager is the project owner's supplier/provider, while the project owner is the project manager's client/purchaser.
- By implication, a project manager is accountable only to the steering committee (through the project owner). The project manager must not be required to report to any other entity in the project governance model, especially reference groups.

Qualifications. A project manager should be strongly committed to the project and have thorough knowledge of project planning and management. The larger the project, the more desirable it is that the project manager have professionally recognised credentials. Like the project owner, a project manager must be politically astute, and capable of managing upwards in his/her dealings with the steering committee, reference groups and advisers.

Engagement. A role definition for the project manager is based on a generic structure that would apply to all projects. The person who assisted the champion to assemble the business case is obviously a strong candidate to fill the role of project manager. This would normally be done as soon as a decision is taken in favour of the business case.

Performance evaluation. The project manager should be assessed by the project owner(s) after a project's outputs are delivered. That assessment should be based on the "steel" tetrahedron (outlined in Sect. 3.2 in Chap. 3) involving criteria for: scope/quality, timeframe, cost and detrimental outcomes.

4.3.4.4 Project Team

Role. The project team builds, assembles, delivers and implements the project's committed outputs. Project teams often demand a formal structure of their own, especially if they are made up of sub-teams and suppliers. Appropriate structures may be based on outputs, skills, phase or even the affiliations of team members.

Members of the project team are accountable to a designated leader/manager for completing assigned tasks (or producing assigned outputs). Project team members may also be assigned specific above-the-line responsibilities (such as managing the risk register). In larger projects, the demands of above-the-line work become so great that it becomes necessary to support the team with specialist project administrators.

Qualifications. Project team members will be appointed according to their capabilities to undertake selected parts of the project's WBS. Project administrators must have a deep knowledge of project planning and management. They would also expected to be persistent and relentless with detail.

Engagement. A role definition is established for each project team member. He/she is engaged by the project manager, or, if the project is large, by an appropriate sub-team leader or contractor. In many projects, key team members are appointed from the ranks of the participating organisations. A number of issues can arise with internal appointments that may require resolution. Two are worthy of particular note. Because staff are effectively "off-line" while working on projects, their career development may suffer, especially if the appointment is long-term and full-time. Part-time appointments bring with them issues of resource contention (because staff now face dual reporting lines). In those cases it may be desirable to cover the assignment with a memorandum of understanding between the staff member, the relevant line manager and the project manager (see Sect. 4.3.2). Such a document would openly summarise the impacts of the appointment on the staff member and map out a plan to address them, as well as establishing ground rules for resource allocation between competing demands.

Performance evaluation. Each team member should be assessed by the manager to whom they report in the team structure using a "local" variant of the steel tetrahedron. Evaluations of those staff drawn from the participating organisations should be provided to HR for incorporation into their personnel records.

4.3.4.5 Reference Groups and Advisers

Role. Reference groups provide specialist input to the project. Reference groups (or "advisers" if a single person is involved) are highly specific and so it is not uncommon to find a large number of these represented in the model (although not all of these may exist at the one time). Reference groups, as well as advisers, support the team in a similar way to subcontractors.

Principles. The decision to locate someone to the reference (rather than the delivery) division of the model is, in general, based on a number of (relatively soft) criteria:

- The first concerns the way the role reports into the governance model. If a role is to report to the steering committee, then it cannot be located in the delivery division. A lawyer commissioned to review the project manager's contract, for example, must be recognised as an adviser.
- The second concerns the nature of the resources provided. If someone is not a "deployable" resource they will probably be recognised as an adviser. For example, if a senior manager from IT is commissioned to confirm the acceptability of a software product selected by the team, then he/she is better viewed as an adviser (even if the role reports to the project manager).

- The third concerns the intensity and duration of involvement. It may not be convenient to have someone who has a relatively minor, periodic involvement with the project assigned to a formal position in the project team structure.
- The fourth relates to the need for independence. Someone who is to comment on (or validate) the work of the team may be better located outside the team.

In some cases, it is an arbitrary decision whether these are located in the reference or the delivery division of the governance model.

Qualifications. Reference group members and advisers will be appointed according to their capabilities to undertake selected parts of the project's WBS.

Engagement. A role definition is established for each reference group member and adviser. Depending on the brief, he/she is engaged by either the project manager or steering committee. Reference group members and advisers can come from inside or outside the performing organisations. The issue of remuneration for reference groups and advisers will be resolved according to the circumstances surrounding each proposed engagement.

Performance evaluation. Where appropriate, the evaluation of reference groups and advisers would normally be undertaken as part of their line roles within their organisation's standing structure.

4.3.4.6 Project Counsellors

Role. Project counsellors ensure that the project is being run in accordance with accepted project and commercial management practice. Projects need not have counsellors, but the larger the exercise, the stronger the case that can be made for appointing them. A project assurance counsellor is normally commissioned by the project owner (or steering committee), sometimes following a recommendation from the project manager.

Project assurance counsellors are expected to:

- Monitor the way that the project's above-the-line work is being carried out.
- Work with the project manager to resolve issues in this area.
- Report periodically to the steering committee.

As well as this general counselling role, on large projects the need may also arise for an additional special role of probity counsellor/adviser. A probity counsellor ensures that the commercial dealings between the project team and the outside world are consistent with the policies, standards, regulations and laws that apply to the project environment.

Not all projects require these roles, it is the steering committee's decision to engage them.

Principles. A project counsellor must not face any conflicts of interest in his/her appointment.

Qualifications. Counsellors are required to have a deep understanding of project management or commercial practice. They would also normally be expected to have well-developed political and organisational skills.

Engagement. Counsellors can come from inside or outside the participating organisations and are appointed by the steering committee (in consultation with the project owner and project manager).

Performance evaluation. The evaluation of reference groups and advisers would normally be undertaken as part of outcomes closeout.

4.3.5 *Managing the Project Governance Model*

The project governance model is assembled progressively over the early phases of a project. During initiation, the project champion (in consultation with the funder and project-manager-designate) designs a broad model by identifying the roles that need to be filled and deciding on the most appropriate assembly of elements corresponding to those roles. While candidates for appointment to certain roles might be known at this time, it may not be possible to complete the model until the planning phase (or even later). There will be occasions where a form of governance is required during the planning of a project (as might happen on very large projects). In those cases, the associated preparatory stakeholder engagement and establishment of the project governance model will all be carried out as an additional set up activity early in planning.

Each specific appointment of an entity to a role in the model should be formalised with a brief of some kind. The project governance model is formally reviewed twice. First, the funder confirms it when accepting the business case. Then any changes are approved later by the steering committee and, where necessary, also ratified by the funder.

The project's key players will normally lead development and operation of the project governance model. This work can be divided into three processes, as is summarised in Table 4.3 and detailed in the following descriptions.

1. *Agree on broad structure of model.* The proposed project model is developed and included in the business case:

   1.1. Review all those stakeholders whose engagement programme includes a role in the project.
   1.2. Assemble a model that recognises all roles required for this project.
   1.3. Decide on the elements required by those roles.
   1.4. Assemble a provisional list of candidate entities for selected elements.
   1.5. Outline all relationships amongst the elements of the model.

2. *Finalise design and assign entities.* Where known, discuss roles and responsibilities in the business case. A fully detailed version of this may only emerge during project planning:

Table 4.3 The project governance model processes

Project governance process	Project phase	Output	Leader	Approved by
1. Agree on broad structure of model	Initiation	Governance model	Champion	Funder
2. Finalise design and assign entities	Planning	Assigned roles in the governance model	Project manager	Project owner
3. Operate/monitor/ maintain	Execution, outcome realisation	A current and appropriate governance model	Project owner	Steering committee

2.1. Prepare generic role definitions:

    2.1.1.   Name of role

    2.1.2.   Division where this role is located

    2.1.3.   Objective of role

    2.1.4.   Desired outcomes of role

    2.1.5.   Outputs from role

    2.1.6.   Core activities and frequencies

    2.1.7.   Membership/leader

    2.1.8.   To whom does this role report (the steering committee, project owner or project manager)?

    2.1.9.   Review of role (who/when/where?)

    2.1.10.   Date of creation

    2.1.11.   Date of termination

2.2. Assign entities to each role.

3. *Operate/monitor/maintain the Project Governance Model.* During the project execution and outcome realisation phases, the project manager continuously updates the project governance model. These processes are also described in Chap. 7 (Execution).

3.1. Induct all entities into their roles.

    3.1.1. Assess capability of each entity to undertake role.

    3.1.2. Design induction programme.

    3.1.3. Induct stakeholder into the model.

    3.1.4. Manage the roles of all project governance model elements throughout their life.

3.2. Periodically review and update the model throughout the project:

    3.2.1. Review operation of the model.

    3.2.2. Review issues/risk registers for project governance model-related entries.

    3.2.3. Propose changes to current model for approval by steering committee.

4.4 The Project Management Office (PMO)

While all projects demand the approval and sponsorship of executive management, some also merit on-going support with additional specialist above-the-line skills and resources. In recent years, many organisations have established a Project Management Office (PMO) to assist those involved in projects with facilities such as programmes of professional development templates, guides and computer tools. Distinct forms of charter for a PMO can be assembled by selecting from the following list of candidate functions:

- *Resourcing.* The provision of specialist above-the-line skills to projects—in particular: planning/management, administration and assurance.
- *Capacity development.* Mentoring, professional development and training programmes for staff and (in larger organisations) sponsorship of communities of practice.
- *Methodological guidance.* Assembly/adoption and implementation of project management frameworks.
- *Standards.* Ensuring adherence to local practice and performance evaluation.

Other responsibilities appear in practice from time-to-time, including portfolio management, project funding and ownership. In light of the principles outlined elsewhere here, roles of that kind lie at the heart of organisational accountability. To assign them outside the senior executive ranks would require such heavy qualification that it is difficult to find a satisfactory rationale for having them filled by a PMO (Box 4.3).

4.5 Stakeholder Management

A stakeholder is an individual or entity who is either potentially impacted by the project or who has a potential impact on the project. Both forms of impact can be positive or negative. For example, a resident of one city who commutes to work in another state could well be positively impacted by a building of a new airport, while a resident of a suburb adjacent to the site might be negatively impacted by the increase in noise from expanded operations at night. Stakeholders are of interest to key players because they can, in some cases exert a significant influence over the way a project unfolds, and, as a consequence over its success or failure. Starting work on a project where powerful stakeholders fundamentally disagree about objectives, strategy, scope and direction is an almost certain recipe for disaster. If issues of that type prove difficult to resolve during initiation, they will become intractable once the project starts. In such a situation, it is better to postpone or abandon cancel the exercise before it begins, rather than during its execution.

Box 4.3 From the Literature

The Project Management Office

The Project Management Institute defines a project management office as follows: "an organisational body or entity assigned various responsibilities related to the centralized and coordinated management of those projects under its domain. The responsibilities of the PMO can range from providing project management support functions to actually being responsible for the direct management of a project" (PMI, 2008). Another definition for the project management office by Ward (2000) is: "A project management office is an organisational entity established to assist project managers, teams and various management levels on strategic matters and functional entities throughout the organisation in implementing project management principles, practices, methodologies, tools and techniques."

Project management offices are created to help in various tasks of project execution and to integrate project related work inside the organisation. Dai and Wells (2004) make a distinction between a project office (engaged to the management if a single project) and project management office (for the management of multiple projects in an organisation).

The experience to date with PMOs is far from encouraging. Hobbs and Aubry (2007) found that they have a remarkably short lifespan with the average being only two years.

Letavec (2006) distinguish between three PMO functions: PMO as (1) a consulting organisation; (2) knowledge organisation, and; (3) standards organisation. Rad and Levin (2002) introduce project-focused and enterprise-oriented functions of a PMO. The project-focused functions include: consult, mentor, and augment. The enterprise-oriented functions include: promote, archive, practice, and train. Hill (2008) introduces five distinctive PMO functions and their subfunctions. Firstly, practice management, including the subfunctions of project management methodology, project tools, standards and metrics, and project knowledge management. Secondly, infrastructure management, including the subfunctions: project governance, assessment, organisation and structure, and facilities and equipment support. Thirdly, resource integration, including the subfunctions of resource management, training and education, career development, and team development. Fourthly, technical support, including subfunctions of mentoring, project planning, project auditing, and project recovery. Fifthly, business alignment, including the sub-functions: project portfolio management, customer relationship management, vendor/contractor relationship management, and business performance management.

4.5.1 The Concept of Project Stakeholding

Stakeholder management often demands additional outputs from the project—beyond those that are implied by the ITO model (See Box 4.4). For this reason, scoping and stakeholder management are intertwined. While a substantive start can be made towards scoping a project before considering its stakeholders, a formal scoping statement cannot be concluded until an initial stakeholder management programme has been formulated (Box 4.5).

Box 4.4 A Case Study: Airport Expansion Project

How Stakeholder Management Influences Project Scope

Consider a project to increase throughput at an international airport. The initial scope was restricted to airport-specific outputs such as a terminal and new runway, but engineering analysis reveals that nearby residents would suffer significant increases in noise. It is decided that, to engage these stakeholders, their homes should be noise-proofed with double-glazed windows and sound insulation. Such a decision would increase the scope of the project (so that "modified houses" are now in scope). This provides another example of the interdependence between different elements of a business case that must be considered during initiation.

Box 4.5 From the Literature

Views of Stakeholders and their Management

Stakeholder management is widely accepted as an important area to enhance project success (Longo & Mura, 2008). We define "stakeholder" very generally—as any individual or entity who has an interest in the project. While other meanings have been suggested for the term, they can all be viewed as broadly consistent with our definition. Some scholars use the organisation's stakeholders as a starting point for a list of those who have an interest in a particular project. According to that approach, stakeholders are any group or individual who can affect or is affected by the achievement of the firm's objectives (Freeman, 1984). In this context, stakeholders are defined as 'narrow' or risk-bearing such as employees, investors, customers, community residents and the environment which may impact a firm (Clarkson, 1995; Mitchell, Agle & Wood, 1997).

Other definitions include "someone affected by a project and having a moral right to influence its outcome" (Bourne & Walker, 2005) and "any group or individual who can affect or is affected by the achievement of the firm's/project's objectives" (El-Gogary, Osman, & El-Diraby, 2006; Freeman, 1984). Stakeholders are entities (individuals, groups, corporations or agencies) who are either potentially affected by the project, or have a potential affect on its conduct and results. Stakeholders may include the funder, project team members, senior managers and employees from other departments of the performing organisation.

Stakeholder management aims at assisting a change leader in identifying, understanding and influencing the impact of external agents on projects (Hillman & Keim, 2001). Because stakeholders have such a significant potential impact (both direct and latent), several management frameworks have been developed (Cleland & Ireland, 2007) and applied in various disciplines, such as e-business (Rodon et al., 2007), managing environmental conflicts (Elias, 2008), software development (Woolridge et al., 2007) and project management (Bourne & Walker, 2005).

Different stakeholders may not only have dissimilar expectations from a project, they may even seek irreconcilable objectives (Mohanty et al., 2005). Hence, it is important that the views of key stakeholders are identified and evaluated to assess potential threats and opportunities for a project (Kerzner, 2009; PMI, 2004).

The adjustments faced by some stakeholders because of change is seen as a generic cause of undesirable outcome from a project, thus the "resistance to change" phenomenon (e.g. El-Gohary et al., 2006; Ford, Ford, & D'Amelio, 2008).

The analysis in the following sections suggests that there is no particular reason that stakeholders who have to adjust to change should fortuitously find themselves as project beneficiaries. In general, a project will involve entities who would be (at least in their own view) better off if the exercise did not proceed and who may, as a consequence, expose the project to additional risk. The accepted response to this problem is framed in terms of "managing change" with a focus on those whose stakeholding is related to their need to adapt to change. The approach outlined below accepts a very general definition of stakeholding, yielding a correspondingly larger and more diverse community of stakeholders. The term "stakeholder engagement" is used here to describe programs that seek to involve stakeholders in ways that increase a project's attractiveness, not only to them, but ultimately to the funder. The objectives of stakeholder engagement will be framed at three distinct levels:

- To increase the support of those who are favourably disposed.
- To decrease the resistance of those who are not favourably disposed.
- To reduce any risks associated with active opposition.

In other words, scope depends on stakeholder engagement, but a comprehensive list of stakeholders cannot be assembled until the project's scope is determined. This linkage between scope and stakeholders is an example of the interdependence amongst elements of a project that can only be satisfactorily addressed with an iterative approach to initiation (for which we propose the initiation/planning spiral).

4.5.2 The Community of Stakeholders

Stakeholders are individuals, organisations or even ad-hoc groupings of people. The forms of impact between project and stakeholder represent the entity's *stakeholding* in the project. In the airport expansion example above, the local residents are stakeholders and their stakeholding can be summarised as:

- Potential impact of the project on the residents: reduced quality of life due to increased noise.
- Potential impact of the residents on the project: generation of a groundswell of opposition to the project.

An entity's stakeholding can be revealed by exploring each of these two forms of impact as questions: "What is the potential impact of the project on the stakeholder?" and "What is the potential impact of the stakeholder on the project?". While in the airport example, the local residents have a stakeholding defined by just two impacts, in general, the stakeholding of a particular entity could be defined by any number of impacts (positive or negative).

Chapter 1 drew attention to the way in which stakeholders add to the management challenges of projects. Indeed, managing stakeholders was found to be one of the most critical areas in project management (e.g. Bourne & Walker, 2005; Crawford, Pollack, & England, 2006).

This section explores the concepts and techniques that underpin such a process. The rationale for including some type of stakeholder management element in frameworks of project management can be found by noting that key players in a typical exercise display wide ranges of characteristics across three areas:

- In the views that they hold about the desirability of the project, ranging, in certain cases, from wholehearted support through to passionate opposition.
- In the power that they can exercise over factors that may impact success.
- In the degree of influence that can be brought to bear on them by proponents of the project.

The first two of these areas (the levels of support from stakeholders and their power to influence the project environment) can be shown as critical to the assessment of a project's attractiveness and so merit inclusion in formal frameworks of project management.

Take two identical projects "A" and "B" and consider a scenario for each, in which there are different patterns of support from one particular stakeholder group,

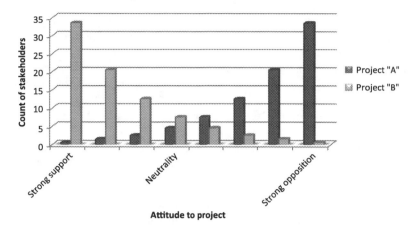

Fig. 4.5 Contrasting distributions of support in two (otherwise) identical projects

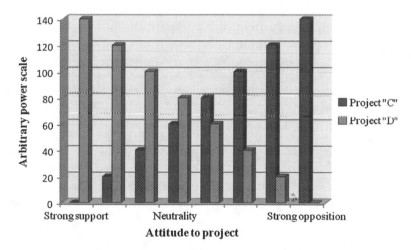

Fig. 4.6 Contrasting distributions of power amongst stakeholders in two (otherwise) identical projects

local residents. "A" has very weak community support, while "B" has very weak community opposition, as displayed by the histogram in Fig. 4.5.

It is clear from Fig. 4.5 that Project "A" is less attractive than project "B" because its lower level of support suggests a higher risk of failure.

In a second scenario, consider two identical projects "C" and "D" where the pattern of support for "C" is identical to the pattern of support for "D". In this case, for ease of discussion we will assume that the numbers of supporters and opponents are equal, resulting in two "flat" versions of the histograms shown in Fig. 4.5. However the levels of power that can be wielded over the project environment by the two communities of stakeholders differ between the two projects. In Project "D", the supporters are the most powerful, while for Project "C" the supporters have less influence. This is shown in Fig. 4.6.

Again it is clear from Fig. 4.6 that Project "C" is less attractive than project "D" because, although numerically the same, its opponents are more powerful than its supporters.

Given the potential impact of these two stakeholder-related variables (distribution of support and distribution of power), it becomes important for a funder to understand three aspects of a project so that a decision can be taken about stakeholder management.

- *Level of support.* What is the nature of the distribution of support for the project amongst its stakeholders?
- *Level of power.* What is the nature of the distribution of power amongst the project's stakeholders?
- *Strategies to manage stakeholders.* What strategies exist to influence both of the above distributions? How effective are those strategies? What costs do the strategies incur?

Consider the following illustration. A project is undertaken to reduce local traffic congestion by extending a freeway, but this involves significant reduction of the local community's access to a nearby botanical gardens. Amongst this community are active supporters of the gardens who: are well-organised, have established a large "fighting fund" and include many people with strong political connections. While (at the moment) they are not opposed to the freeway per se, they find it unacceptable that the gardens will be effectively isolated from most local roads. At the same time (although this is not yet known to the project proponents), a particular modification to the freeway design would alleviate the bulk of the community's concerns. Furthermore, the costs of such a change would still leave the project highly attractive to the funder. In such a situation, it is clear that the project proponents would be well-advised to:

- Recognise the residents as stakeholders in the project.
- Understand the nature of their stakeholding.
- Set a strategy and establish a plan for dealing with their concerns (which, in this case involves a change in project scope).

These steps offer an insight into one of the core elements of a project management framework—stakeholder management. Stakeholder management can be viewed as a highly evolved form of risk management which mitigates the risk of damage to the project arising from stakeholders becoming "detached" from the initiative. As is the case with risk management itself, stakeholder management involves the funder in a trade-off between risk and cost whereby additional outlays are accepted so that the likelihood of a successful project is increased.

4.5.3 The Stakeholder Management Process

The overall stakeholder management process is triggered by a number of scenarios, the most obvious of which is the preparation of a business case. The process will also be executed again during planning when a stakeholder engagement plan is mapped out in detail. Other triggers relate to a project that is under way and for which changes arise in the project's scope, the community of stakeholders or the interests of existing stakeholders. As was suggested in the background to the freeway project above, the management of stakeholders involves five processes, presented in Table 4.4.

Stakeholder management is ongoing throughout the entire project, starting very early during the initiation phase, when key stakeholders are first identified and analysed. Both these processes are important precursors to the assembly of an appropriate programme of engagement for key stakeholders in the project. This is usually done by analysing the stakeholding of those who have an interest in the project and gauging their potential support or opposition. Stakeholder analysis goes beyond an exploration of the interests that various entities have in the project, it also includes the development of an outline strategy for their engagement. A strategy for the engagement of a particular stakeholder is simply a list of the agreed actions that are to be taken to get the entity involved in an appropriate way, but it contains no details about those actions. For example it might be decided to engage a community action group by conducting a series of public meetings. At this point there is no detail about the proposed meetings (such as the format, venue or conduct) and so the next step in stakeholder engagement assembles a detailed plan for the strategy. The project manager normally administers the process, but much of the plan itself involves the project owner and, in due course, other members of the steering committee. Substantive work on implementing the engagement plan begins during execution. During execution, the project manager

Table 4.4 The project stakeholder management processes

Project stakeholder management process	Project phase	Outputs	Leader	Approved by
Identification	Initiation (ongoing)	A list of stakeholders	Champion	Funder
Analysis	Initiation (ongoing)	Stakeholder register	Champion/ project manager	Project owner
Engagement planning	Planning (ongoing)	Stakeholder engagement plan	Project manager	Project owner
Engagement implementation	Execution (ongoing)	Engaged stakeholders	Project manager	Project owner
Engagement monitoring	Execution (ongoing)	Updated stakeholder engagement plan	Project manager	Project owner
		Updated stakeholder register	Project manager	Project owner

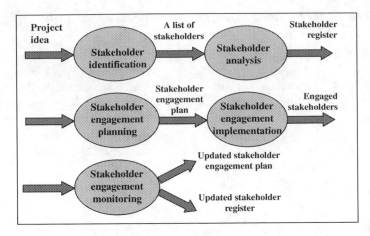

Fig. 4.7 Stakeholder management processes

Fig. 4.8 The stakeholders at Project BuyRite

will also monitor the effectiveness of the stakeholder engagement plan and seek to control its operation.

In summary, stakeholder management continues throughout the entire life of the project, based on the processes presented in Fig. 4.7.

4.5.4 Stakeholder Identification

The first stakeholder management process involves the identification of all project stakeholders. It is important that all entities who have an interest in the project be identified in this step, especially those who are negatively impacted by the exercise. Identifying project stakeholders can be assisted with knowledge from previous projects and from the experience of the funder and the promoter. Some of the stakeholders in Project BuyRite, for example, are shown in Fig. 4.8.

In general, stakeholders emerge in a project for one of two reasons: an impact between the project and an entity arises spontaneously or an impact arises because the entity is assigned a role in the project. Take the freeway example above. Supporters of the botanical gardens are stakeholders regardless of any roles that might be recognised in any (eventual) governance model for the project. By way of contrast, the person who is eventually appointed as project manager becomes a stakeholder solely by virtue of that appointment. (The special case of the project manager who also happens to be a supporter of the botanical gardens has no effect on this general distinction).

Spontaneous impacts can arise at any point in the life of a project. In the case of the freeway:

- Those householders whose properties adjoin the construction site and who will suffer from dust and noise are spontaneous stakeholders because of the (negative) impact on them arising from the work of constructing the freeway.
- Those motorists who use the completed freeway will experience reduced travel time, a (positive) impact associated with achievement of a (presumed) target outcome. These motorists are also spontaneous stakeholders.
- The Botanical Gardens Trust (as an organisation) is a spontaneous stakeholder because of a negative outcome. If the freeway is completed to its original design, then patronage will fall because of reduced access to the site.

When identifying stakeholders, the primary focus is on these "spontaneous" stakeholders because they are imposed on the project by circumstance. There are six generic classes of spontaneous stakeholders in a project:

1. *Funders.* Those who have discretionary authority over the resources that will be released to the project (The state government).
2. *Beneficiaries.* Those entities who are targeted by the funder to receive a "flow of value" from achievement of desirable outcomes from the project. (Motorists).
3. *(Positive) Impactees.* Those who experience a fortuitous "gain in value" from the project unrelated to its target outcomes. Take the case of a mobile canteen based in the general vicinity of the freeway construction site. The operator of the service may well become a positive impactee because of increased sales to the workers. (The operator is not, however, a beneficiary because the funder is not seeking an increase in canteen sales as a return on the investment on the freeway). (Local businesses).
4. *(Negative) Impactees.* Those who experience a "loss of value" from the project. (Neighbours).
5. *Customers.* Those who will utilise the project's outputs and, in so doing, generate target outcomes. (Motorists again).
6. *Influencers.* Those who, by virtue of their position, or standing are able to carry a significant body of opinion about the project. (The local newspaper).

Figure 4.9 positions some of these stakeholders in relation to the ITO model, for example, the funder providing the funds and resources (inputs) to the project, or the project manager leading the process.

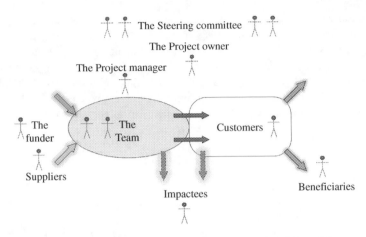

Fig. 4.9 A selection of stakeholders positioned with respect to the ITO model

Note that in particular cases, a specific entity may be a member of any subset of these classes. Take for example, a resident who lives next to the proposed construction site and who will also have travel time reduced when the freeway is completed. That person is both an impactee and beneficiary.

Later, as the initiation spiral progresses, work on the *stakeholder* and *governance* elements will result in other entities becoming "commissioned" stakeholders because of their appointment to fill specific roles in the project. There are nine generic classes of commissioned stakeholders in a project:

1. *Suppliers.* Those who provide goods/services to the project. Suppliers can be internal, external, commercial or non-commercial.
2. *Project owner.* A person who is held accountable by a funder for the realisation of target outcomes.
3. *Champion.* The person who drives preparation of the business case. The project champion is obviously a candidate for eventual appointment by the Funder as project owner.
4. *Team members.* Those engaged in the work of: producing project outputs, planning the exercise, managing that work or administering it.
5. *The project manager.* The person held accountable by the project owner for delivery of the project's outputs.
6. *Steering Committee members.* The steering committee is a small group of powerful project supporters who oversee the project's execution and eventually outcome realisation. More is said of this group in Sect. 4.3 (project governance).
7. *Reference group members.* Reference groups are created to provide specialist (collective) input to the project.
8. *Advisers.* Appointed to provide specialist (individual) input to the project.
9. *Counsellors.* Specialists who are appointed by the funder, owner or steering committee to ensure that the project is being run in accordance with accepted practice.

Appointments using the list above are obviously constrained by what has been revealed using the list of spontaneous stakeholders. These classifications are useful because one can postulate modes of behaviour for each class that allow predictions to be made about the response of particular entities to different project scenarios. For example, a negative impactee who has indicated implacable opposition to the project and who has become totally committed to its failure would be disqualified from joining a steering committee. Such predictions can then help in the formulation of engagement plans. Amongst other things, stakeholder analysis determines to which of these classes each stakeholder belongs (or will eventually belong when roles have been assigned).

It is inevitable that a particular entity may hold multiple forms of stakeholding in a project. For example, it is common for the funder to be a beneficiary and an influencer. It is also worth noting that some misunderstandings about the nature of project stakeholding have given rise to flawed practices. For example, appointing an external supplier to a steering committee places that person in a position of conflict of interest and so must be avoided.

4.5.5 Stakeholder Analysis

Analysis is concerned with understanding the nature of each entity's stakeholding in the project so that their current position (in terms of goals, attitudes, power and behaviours) can be assessed. It is also concerned with the assembly of a strategy for the engagement of selected entities. Engagement is defined as a managed relationship through which the key players seek to influence other stakeholders in a way that will contribute to the success of a project.

4.5.5.1 Exploration of Stakeholding

Two pieces of information are central to understanding an entity's stakeholding in a project:

- *Nature of stakeholding.* This describes the nature of the interests that an entity has in the project.
- *Classes of "spontaneous" stakeholding.* This is a list drawn from the six classes shown before.

Stakeholder analysis will often reveal contradictory and conflicting positions among stakeholders. Take for example, a project to increase visitors to the historic precinct of a provincial city and consider a café in that precinct. If the target outcome is increased revenue for local businesses then the owner of the café is a beneficiary. But if, at the same time, the resulting congestion lowers the quality of life for residents, then they are disbeneficiaries. A funder must then make a trade-off amongst the conflicting interests of these stakeholders.

4.5.5.2 Issues Arising from the Entity's "Spontaneous" Stakeholding

In general, there will be no issues arising from someone's stakeholding (other than the obvious ones, such as a negative impactee being unhappy about the project and likely to foment opposition). Based on this, together with opportunities to influence each stakeholder, the project proponents can determine which forms of intervention are desired and feasible.

4.5.5.3 Strategy Formulation

Based on what has been discovered during the exploration of each entity's interest in the project, a decision can now be taken on how to engage those who need to be involved.

While engagement analysis is often undertaken without the direct participation of stakeholder concerned, in particular cases it may also be appropriate to involve them directly. Face-to-face discussion during analysis may encourage stakeholders to actively and openly discuss their attitudes towards the project, allowing more effective programmes of engagement to be formulated.

It is here that the information assembled to date in the stakeholder register serves a critical purpose, by stating clearly what the purpose is of engaging this stakeholder (thus helping to frame an appropriate and achievable programme of involvement. The output of this process takes the form of a stakeholder engagement strategy acceptable to the funder.

A useful structure for this recognises three generic forms of stakeholder engagement:

1. *Include the stakeholder in the project's communications plan* An example would be to invite the stakeholder to a regular monthly briefing by the project manager. A corresponding entry in the stakeholder register would take the form of: the phrase "Include in the communications plan" followed by identification of the proposed forms of communication such as "website access" or "regular newsletter".
2. *Make the stakeholders the subject of a programme of special engagement* When a large steel mill closed some years ago in Australia, those staff who were to lose jobs became stakeholders in the exercise. In addition to including them in the project's communications plan, the Company offered each person a special "Pathways" programme to help them manage the changes they faced. For example, some were guided into the establishment of their own business. Programmes of special engagement have the potential to significantly expand a project's scope. A corresponding entry in the stakeholder register would take the form of: the phrase "Include in programme of special engagement" followed by identification of the proposed programme such as "Provide temporary accommodation during earthmoving" or "offer redundancy package".

3. *Include this stakeholder in the project governance model* A project to move office may have important implications for staff travel. In that case, it may be helpful to form a reference group of staff to offer comments and provide input to the project team. A corresponding entry in the stakeholder register would take the form of: the phrase "Include in the project governance model" followed by identification of the proposed role such as "include in community liaison reference group" or "invite as member of technical standards working group".

For example in the freeway illustration, the engagement strategy for local residents may appear as:

1. *Include the stakeholder in the project's communications plan.*
   - Provide background material on the project website.
   - Conduct monthly public briefings.
   - Include in the distribution list for the project newsletter.
2. *Make the stakeholders the subject of a programme of special engagement.* Design a programme of special engagement by including an on/off-ramp to the botanical gardens:
3. *Include this stakeholder in the project governance model.* Create a reference group made up of representatives of the botanical garden supporters.

Face-to-face contact may figure prominently on the strategy assembled for particular stakeholders, but care has to be exercised when considering such an approach, especially with entities who are opposed to the project. Of particular note are situations related to "community action" and "special interest" groups where public meetings may play a critical role. Activities of this kind are challenging, but if done well, can go a long way towards ameliorating the worst effects of public antipathy. The consequences of poorly conducted public meetings can be catastrophic and so they must be planned, promoted, structured, and chaired thoroughly and professionally.

4.5.6 The Stakeholder Register

The primary tool in stakeholder management is the stakeholder register, which is used to document the results of analysis and outline the form of engagement being proposed for each project stakeholder. The stakeholder register is held in the business case and later in the project plan. Because it is a register, this tool takes the form of a table where rows are associated with instances (stakeholders) and columns with attributes. A set of attributes making up a typical stakeholder register would include columns for:

1. *Entity name.* The name of the entity being considered as a stakeholder.
2. *Stakeholding.* Listing the impacts of the project on the entity and the impacts of the entity on the project.

3. *Classes of "spontaneous stakeholding"*. Listing the classes (as identified in Sect. 4.5) to which this stakeholder belongs.
4. *Issues arising from this stakeholding*. Listing issues associated with this entity's stakeholding that need resolution.
5. *Objective of engagement*. A clear statement of what results (outcomes) are expected from engagement of this entity.
6. *Engagement strategy*. Identifies the mechanisms that will be employed to engage this entity appropriately in the project.
7. *Commissioned stakeholding*. Identifies appropriate assigned roles (if any) for this entity in the project governance model. (Note that this ignores those roles filled by people who are not spontaneous stakeholders).

The project manager holds the primary stakeholder register, but there may be occasions when others hold "supplementary" registers. For example, the project assurance counsellor (when the role exists) may assemble a confidential stakeholder register when premature release of information about an opponent would cause irreparable damage.

4.5.7 Stakeholder Engagement Planning

With the engagement strategy in place (and summarised in the stakeholder register), the detailed planning of stakeholder engagement can now proceed, as part of the "standard" work undertaken during the global project phase called planning. Details about all the work involved in stakeholder engagement, and in producing the outputs associated with stakeholder engagement (such as websites, newsletters and so on) are incorporated into the normal planning activity and eventually reflected in the work breakdown structure, Gantt chart and schedule of milestones.

When addressing that part of the engagement strategy that relates to communications plan, it is necessary to integrate, harmonise and reconcile all the separate elements into a cohesive communications plan. Fragmented, conflicting and uncoordinated attempts at communication could have the opposite effects to those established in the original engagement strategy.

4.5.8 Stakeholder Engagement Implementation

Implementation of the stakeholder engagement plan is broken into two parts: some completed (perhaps surprisingly) during planning and the bulk of it undertaken (as expected) during execution, and much of that very early in the phase. The foundation for this work is the engagement plan. This provides three areas of detail to guide implementation:

1. *The communications plan.* Which will specify what forms of communication are to be undertaken. For example, if a website is identified in the communications plan, then its detailed design, development and operation will be done in this phase.
2. *Programmes of special engagement.* Which are absolutely peculiar to the project in hand. The work involved in each will have been detailed during planning and is now undertaken during execution.
3. *Involving stakeholders in the conduct of the project.* The situation with implementation of the project governance model is slightly different. The project governance model is established during set-up (which is part of the planning phase). As execution progresses, work on the project governance model is primarily concerned with managing any changes that prove necessary.

4.5.9 Stakeholder Engagement Monitoring

In order to ensure the stakeholder engagement plan is implemented successfully, the project manager has to monitor its effectiveness together with other controlling processes that he/she does during project execution and outcome realisation phases.

The case study in Box 4.6 (Elias and Zwikael 2007) demonstrates how the suggested process has been implemented in a project to engage opposing stakeholders back into the project.

4.6 The Programme Environment

A programme is a collection of projects that are linked by a framework of coordination. Accordingly, the programme environment can be viewed as an extension of the project environment. It is important to note that, by definition, projects have target outcomes. It will become clear in the following discussion that programmes also have outcomes (because they "inherit" the outcomes of their component projects).

Programmes emerge from decisions to link related projects by coordinating their planning and management. Three stylised forms of coordination are introduced in the following pages, distinguished by the level of communication and harmonisation required. Projects are linked into programmes through their governance models.

Before discussing the issue of how a collection of projects might be linked, we need to examine the opposite question: "Under what conditions is it meaningful to divide a project into a set of (smaller) projects"? The answer involves careful

Box 4.6 Case Study: Stakeholder Management in a Construction Project

How Stakeholders can be Engaged Effectively into the Project

This case study analyses a road construction project, managed by the Greater Wellington Regional Council (GWRC) in New Zealand (Elias & Zwikael, 2007). This case study represents a proactive approach to the model described above, where key stakeholders take an active part of the analysis process.

The GWRC had been seeking to address the increasing problems of congestion, safety and community severance along the existing State highway route north of Wellington. The Western Corridor carries from 20,000 to 75,000 vehicles and 11,500 rail passengers per day. According to the GWRC, there was an urgent need of an affordable, safe, efficient, reliable and sustainable Western Corridor transportation network that provides reasonable capacity. In this context, the GWRC proposed a Western Corridor Project, which included public transport improvements, travel demand management initiatives and a staged programme of road improvements that address safety, reliability and capacity.

Some stakeholders supported the project while others voiced their opposition towards it. The conflicting views among stakeholders presented increasing challenges to the transport managers of GWRC. In response, a stakeholder engagement plan was assembled. These stakeholders were successfully engaged in a process based on the following steps.

Those with an interest in the Western Corridor project were identified by GWRC managers. They include internal stakeholders (such as the Finance Department, and GWRC transportation managers), as well as external stakeholders (such as the government and 'green' representatives).

Representatives from all stakeholder groups were invited to participate in a 'raising issues' session. The stakeholders who attended the session generated a total of 72 issues, each related to an opportunity or obstacle, based on their opinions on the project, such as 'cost of accidents', 'alternative use of money' and 'environmental damage'. The stakeholders then grouped issues that were related to form clusters. A descriptive name has been given to each cluster. In this exercise, the stakeholders developed 12 such clusters. For example, the 'cost' cluster involved eleven out of the 72 issues, related to aspects such as the high cost of the project, sources of funded, alternative use of the money and cost of accidents. Stakeholders then identified variables associated with each of these clusters. For example, amongst the 21 variables included in the 'cost' cluster, were 'Cost of congestion', 'accidents', 'number of days road is closed due to hazards' and 'travel time'. All variables were then linked to generate a 'causal loop model'. For this purpose, stakeholders tried to establish relationships between all variables. They identified pairs of related variables—connected by an arrow.

In this process, stakeholders were able to analyse the relationships among the following variables: 'average number of trips per person per day', 'accidents', 'number of days the road is closed due to hazards', 'social impact on community', 'support of community stakeholders', 'support of political stakeholders', 'Western Corridor project', and 'change in trip volume and 'distribution'. At the end of this exercise, all agreed that this model represented a shared view of the exercise. Understanding the interest of different stakeholders in this project, and its potential impact on each of them also contributed significantly to the development of the project business case.

adherence to the definition of a project. A project can be broken up into smaller projects only if each component has a valid scoping statement (with a statement of objective, target outcomes and committed outputs). In other words: if an initiative has outputs, but cannot generate target outcomes on its own, then (if it is to be undertaken at all) it can only represent a component of a (larger) project—it cannot be a project in its own right. Now having said that, we quickly run into a problem of terminology.

Consider a project to reduce the costs of access for heavy trucks to the general cargo area of a large seaport involving: a new bridge over an estuary, a widened road and modifications to a number of overpasses (to increase their clearance). In scenario "A", access costs will not change unless all three outputs are delivered, while in scenario "B" delivery of each output would be associated with incremental decreases in access costs. Under scenario "B" the initiative could, if desired, be broken into three projects (each supported by a separate business case). In fact, in that situation, it may even make sense to fund only one or two of the individual projects. Under scenario "A" however, the exercise cannot be broken up in that way and so a single business case must embrace all three outputs. Despite this, the business case for scenario "A" imposes no particular constraints on when each output must be delivered. It could well be that scarcity of funds or resources force the proponents into a strategy of progressive delivery over many years. It is here that we face a terminological issue. If the overpasses are to be modified in year 1, the bridge erected in year 3 and the road widened in year 6 (with lengthy periods of inactivity between them), how are we to refer to the three individual exercises—and to the overall initiative of which they are a part? While the formality of the approach presented here requires that the work be viewed as a *project* made up of three *sub-projects*, we do acknowledge that such terminology sits awkwardly with more common parlance (in which it may well be described as a "*programme* of *three projects*").

4.6.1 Related Projects

A relationship amongst projects will emerge in either of two situations:

1. *Imposed dependency.* Where it is discovered that the worth of one project will be reduced because its execution is constrained by the execution of another project. This sort of relationship has the effect of increasing duration or cost (or both), relative to what would have happened without the dependency. For example, the duplication of a rail line between a mine and a port may be delayed by urgent flood mitigation work on a nearby river. It is clear that in order to prevent unnecessary and avoidable loss of worth (due to cost and time increases), both projects should be scheduled and monitored in a coordinated way.
2. *Elected dependency.* Where it is discovered that the worth of one project could be increased by voluntarily coordinating its execution with the execution of another. Consider two (currently) unrelated projects: one to reduce groundwater pollution by connecting all houses in a suburb to a new sewage treatment plant and another to increase community access to a range of services by running a fibre-based high-speed broadband network to those same houses. It may be possible to significantly increase the worth of one or other of these projects by integrating their planning and management, so that costs and durations (or both) can be reduced. For example, the trench for the sewer pipework may be modified to accept a fibre optic cable.

Both forms of dependency can arise spontaneously (as might happen when the projects are undertaken by different funders) or intentionally (when a collection of related projects is purposely assembled by a single funder, perhaps to give effect to a strategy or policy). Consider strategy by an IT company to reduce its reliance on manufacturing by moving into services (such as IT outsourcing and consulting). To achieve its objective, the firm may well commit to a programme of related initiatives including:

- Mergers and acquisitions
- Service development
- Staff development
- Skill acquisition
- Sales force redeployment
- Asset disposal

Some of the items in this list may make sense as projects in their own right while others may, in fact, represent groups of related (and yet-to-be-articulated) projects. Eventually as the component projects are defined, any dependencies amongst them will become apparent, requiring coordination.

4.6.2 Coordinated Projects

Regardless of whether the relationships between projects are fortuitous or intentional, the coordination options are the same:

1. *Loosely coordinated projects.* Appoint a coordination reference advisor within each project, filled by a representative from the other related projects, to monitor progress and report back (so that the plan for each project can be adjusted as work on the other one unfolds). The representative could be a team member, an administrator or even the project manager. The adviser would meet regularly with relevant project managers or project administrators providing information about developments with the "other" project. If the adviser is the "remote" project manager, then he/she may also report to the "local" steering committee. Clearly, this approach becomes unwieldy as the number of related projects grows (and, in fact, may prove impractical with more than two or three). An example of a governance model of a loosely coordinated programme is shown in Fig. 4.10.
2. *Tightly coordinated projects.* Establish a (single) common coordination working party (within the reference division of all related project governance models) to consider planning options for each project and make recommendations accordingly. An example of a governance model of a tightly coordinated programme is shown in Fig. 4.11.
3. *Integrated projects.* Place the related projects under the control of one steering committee and one project manager, with sub-project managers as required. An example of a governance model of an integrated projects programme is shown in Fig. 4.12.

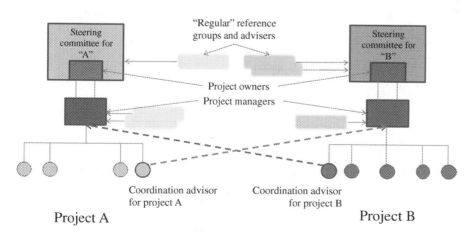

Fig. 4.10 A programme of *loosely coordinated* projects (with independent coordinating reference groups)

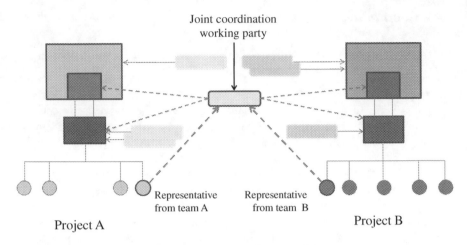

Fig. 4.11 A programme of *tightly coordinated* projects (with a joint coordinating working party)

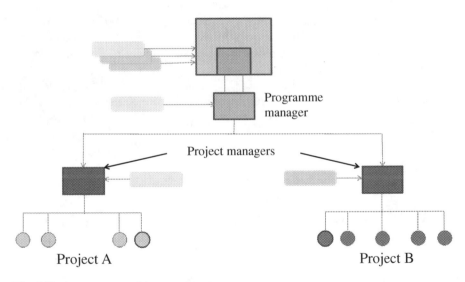

Fig. 4.12 A programme of integrated projects

It will be obvious that as we move from the first to the third option, the strength (and presumably the effectiveness) of the associated linkage is increasing, however, so too is the severity of the constraints on each project. It should also be noted that the prospects for coordination are heavily influenced by any relationships amongst the funders. Disparate funding make projects much more difficult to coordinate than funding from a common source. For this reason, the third option would require that a single funding source be established—if necessary by creating a special commercial entity to act as the funder. Thus, the third option would be open to the IT company mentioned above that has proposed a strategic move out of manufacturing into services.

The discussion of coordinated projects can be summarised thus:

- A programme is a collection of projects whose execution is coordinated.
- Projects are coordinated through their governance models.
- Projects may be linked into programmes for either of two reasons: they interfere with each other (because of an imposed dependency) or their worth can be increased by coordinating their execution.
- Projects become candidates for linkage fortuitously or purposely (when they form part of a coherent strategy).
- Three models of coordination are consistent with the principles of governance proposed here: loose coordination (through cross-project representation), tight coordination (through unified planning) and project integration.

4.6.3 Staged Projects

A special class of programme is worthy of mention. Research and development projects typically involve a sequence of serially dependent stages. The progress of a new drug from research lab to marketplace is a typical example. Each stage, such as obtaining Food and Drug Administration (FDA) approval, is a project in its own right. Except for the last one, each stage has a binary target outcome—along the lines of "The Company placed in the position of being able to make a reliable decision about funding the next stage". (The output from this stage would take the form of documents, certificates and reports). Clearly the overall programme is driven by the target outcomes attached to the last stage which could include "increased revenue", "increased market share" and so on. If, for example, FDA approval is not obtained, then while the Company has certainly achieved the target outcome set for this stage (of making a reliable decision to abandon the exercise), it is denied achievement of any objectives it might have set for the final stage.

The decision to coordinate a group of projects as a programme is taken very early in initiation. From that point on, the entire programme accords, in general, with the frameworks proposed here. Loosely and tightly coordinated projects are all run separately, but constrained by the demands of coordination.

Integrated projects are obviously also run as projects, except for the substitution (where appropriate) of the word "programme" for "project". So, for example in the governance model there would be a programme owner, a programme manager (to whom would report "project managers" rather than "sub-project managers"). In the case of a sequentially staged project, aside from an early overarching document describing the programme, it would be conducted as a succession of independent projects.

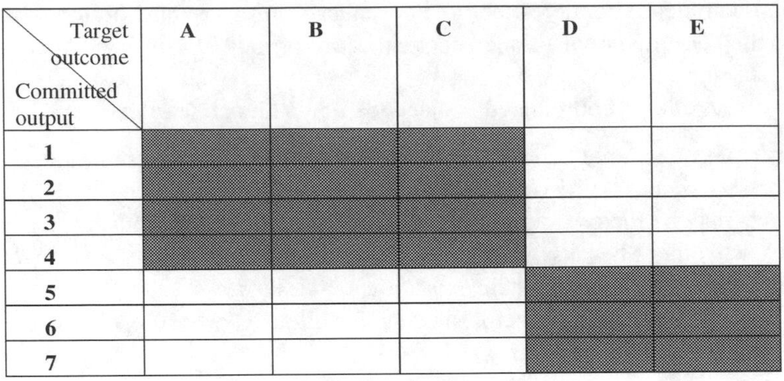

Target outcome / Committed output	A	B	C	D	E
1	▨	▨	▨		
2	▨	▨	▨		
3	▨	▨	▨		
4	▨	▨	▨		
5				▨	▨
6				▨	▨
7				▨	▨

Fig. 4.13 The utilisation map implementation for project partitioning

4.6.4 Partitioning Projects

There will be occasions where an exercise that has been proposed as a single project can actually be split into independent components, so that a separate decision can be taken about an appropriate form of linkage (if any). Fig. 4.13 shows how the utilisation map can be used to split what was proposed as a single exercise into two independent projects based on shared outputs and outcomes. A full description of the utilisation map can be found in Sect. 5.2.5.

In this case the "project' is really two independent projects. The first involves outcomes A, B and C with outputs 1, 2, 3 and 4. The second involves outcomes D and E with outputs 5, 6 and 7. (It is, of course, a separate issue whether or not they should be linked into a programme for management purposes). The funder has considerably more flexibility with independent projects than is the case when they make up one integrated exercise.

4.7 The Project Portfolio

An emerging popular view is that, in some sense, projects/programmes/portfolios represent a three-level scale of desirability or maturity. In the discussion above we have suggested that programmes are little more than coordinated projects, but the concept of portfolios bears some further exploration.

We define a portfolio as simply the collection of projects that the organisation has accepted for funding. All organisations then, face a fundamental portfolio problem which can be expressed as a question:

Which of a proposed list of candidate projects should be funded? To answer this question, the funding organisation must:

1. Classify all of the set of all candidate projects as suitable/unsuitable for funding (thus creating a dichotomy).

2. Rank all items from the "suitable" set in terms of their attractiveness. (Attractiveness has been defined in terms of worth and riskiness).

It is clear that the selection will be constrained by the funds to which the organisation has access. If the combined outlays required by the suitable set fall short of the available funds, then all of them will be accepted, otherwise only some of the suitable set will be funded. In that latter situation, the selection of the projects to be funded is not a trivial matter because projects are "lumpy". In some situations it could well be better to fund two low-ranked candidates than one high-ranking alternative. Techniques have been developed by the Operations Research discipline for making such selections, but they lie beyond the scope of this discussion.

4.8 Summary

In this chapter, we have examined four critical features of the project environment by considering: the way it evolves in the course of a project, the elements that give it structure, the engagement of key players and their organisation. We also explored the concept of a programme, and noted that programmes are handled in a fashion very similar to that suggested for projects. We are now in a position to discuss in depth project initiation, the first stage on the life of a project.

Chapter 5
Starting a New Project

5.1 Initiating a Project

Projects start with an idea for desired change that is seen as beneficial in some way. The initiation phase provides a formal approach to the exploration and analysis of the idea behind a project so that a confident decision can be made about funding it. Because money and resources are allocated to a venture today in the anticipation of a flow of beneficial effects (target outcomes) over some future timeframe, a project represents a form of investment. Consequently, initiation considers (in considerable depth) the outcomes sought by the funder as a return on this investment.

5.1.1 The Project Champion

Initiation involves two recognisable roles: the first associated with conceiving of the original idea for the exercise, and the second associated with formalising that idea as a business case. We identify the person (or entity) filling the first of these as the project promoter and the second the project champion. Occasionally, both roles are filled by the same person.

The project champion is responsible for preparing and tabling the business case and, accordingly, drives and guides the process of initiation. A project champion is appointed as someone who can secure support from a potential funder to investigate the idea further. If successful in this, the project champion will usually be made accountable by the funder for development of a business case. In most cases, the project champion will not have the skills and capabilities (or the time) to write such a document, and so he/she will appoint someone else to do this. The person writing the business case should be skilled in project management and may later become the project manager. It is not recommended that the champion assumes the

O. Zwikael and J. Smyrk, *Project Management for the Creation of Organisational Value*, DOI: 10.1007/978-1-84996-516-3_5, © Springer-Verlag London Limited 2011

role of project manager. One of the reasons relates to the different capabilities and personal characteristics required to promote a project when compared with those required to manage it. It has been found that project champions usually prefer to work alone and are often technologically inclined whereas a typical project manager is more comfortable working with groups, managing people and taking on managerial responsibilities (Kerzner, 2009). The champion and the person writing the business case, together with others who can provide appropriate support, are identified in the following discussion as the "initiation team".

Some additional principles should be borne in mind when appointing a project champion:

1. *Champion as supporter.* The champion should be a strong supporter of the project.
2. *Separation of funder and champion.* Except in small projects, the project's funder should not act as champion. This reduces the possibility of conflicts of interest that might arise if the project is approved by the same person who promotes it. This separation also encourages rigour and formality in the analysis on which the funding decision will eventually be based.
3. *Champion as project owner.* When the business case is eventually accepted, the champion is a strong candidate to fill the role of project owner and hence remain involved throughout execution.
4. *Champion owns the business case.* Even though, for a variety of reasons, the champion arranges for someone else to assemble the business case on his/her behalf, it remains the champion's document. (It does not belong to the person assembling it). This principle has quite profound implications for the handling of the business case, as suggested in Box 5.1.
5. *Qualifications for championship.* The champion should have a good understanding of the business, as well as a working knowledge of project management.

5.1.2 Leading Project Initiation

The initiation phase involves two core processes. The first one is conceptualisation, in which the initiation team briefly investigates the attractiveness and feasibility of an idea as a project. The second is business case development, also referred-to here as "project specification". We use the word "specification" in a variety of ways and care has to be exercised (particularly in the following discussion) to distinguish between them. For example, "project specification" relates to a business case, whereas "output specification" relates to the technical definition of an output identified in the business case. Conceptualisation ends with a decision by the funder to devote further time and resources to the development of a business case. If, at this point, the idea still has the support of the funder, it progresses to become the subject of a formal project specification. When

Box 5.1 Implications for Project Practice

Who "signs-off" the Business Case?

Conventional practice requires that the person actually assembling the business case must get the project champion to "sign-off" the document. There is a range of interpretations of what such a step actually means. The most innocuous is that the champion is merely taking delivery of a document that, at least for the moment, can be treated as complete. A more problematic interpretation is that the champion is accepting a commission by the author of the document to fill that role (thus completely confusing the respective roles of client and supplier).

We suggest a more complete (and less ambiguous) protocol:

- The champion commissions someone (the "author") to assemble the business case. The author accepts a brief for this role. The champion is the client and the author is the supplier in this agreement.
- When the business case is complete, the champion takes delivery of it. At this point the author has fulfilled his/her commission.
- When the project is approved, the funder accepts the business case (by "signing it off") and commissions a project owner (usually the champion).
- The project owner now commissions a project manager (frequently the author of the business case) to prepare a project plan.

In this approach, the term "sign off" should be restricted to two situations: the acceptance of the business case by the funder and the acceptance of a commission by anyone who is invited to fill a formal role in the project.

specifying a project, the initiation team considers (amongst other things) the project's objective, outcomes, outputs, risks, stakeholders, governance and constraints. The result of this work is incorporated into a business case, which is then presented to the project funder for a decision about proceeding further (to the planning phase). Figure 5.1 outlines the project initiation processes and their outputs.

In summary, four clear key steps emerge during initiation, as described in Table 5.1

It should be noted that the *accountable person* is not necessarily the *person doing the work*. For example, the business case will often be assembled by the project-manager-designate (on behalf of the champion).

The following sections explore in detail the two initiation processes: conceptualisation and business case development.

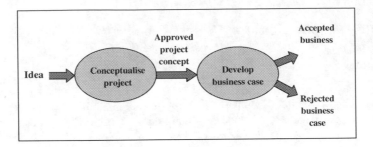

Fig. 5.1 The project initiation processes

Table 5.1 Project initiation steps

Related process	Initiation step	Accountable person	Output
Conceptualise project	Propose the original idea	Project Champion	Project concept
	Approve the project concept	Project Funder	Approved project concept
Develop business case	Assemble the business case	Project Champion	Business case
	Accept/reject the business case	Project Funder	Accepted/rejected business case

5.1.3 Conceptualisation

Discussion about a project can arise from almost any quarter, including a funder, a business unit manager, a services department, an employee, an external consultant or even a supplier. For example, a project could be initiated with an idea raised by the marketing manager, who sees an opportunity for a new product that will fit the unique requirements of the Asian market. Other projects may start because of a constraint, such as new legislation requiring a reduction in the greenhouse gas emissions produced by an organisation. In that case, the project promoter could be the organisation's legal advisor who recognises the need for a response to a new law.

Conceptualisation is the initial investigation into such an idea. To be considered as a project, each idea should be formulated as an investment opportunity. Conceptualisation is intended to convince a prospective funder that an idea merits recognition as a potential project. If so, a business case will then be developed during, what is a relatively informal process, often undertaken with few dedicated resources. Approval of the project concept does not necessarily imply that a future business case will be accepted, it only means that the proposed exercise appears worthy of further investigation. Conceptualisation should be a short, simple and relatively spontaneous process, so that enthusiasm and momentum can be maintained.

Frequently, the idea that triggers this process is initially framed in terms of outputs (rather than outcomes). This is a natural and desirable situation because people often find it easier to think about "things" than about "effects". For example, an employee sees a product at a trade show and suggests buying it, usually because the outcomes are implied (and, to begin with, unstated). Conceptualisation seeks not only to nurture the underlying idea but also to have key stakeholders ask "why?", because at this point the rationale for the proposal is usually far from clear. Thinking about "why" will lay the foundation for a much more formal treatment of target outcomes later.

Most ideas do not evolve into projects. Even good ideas are frequently rejected from further consideration for many reasons that might include:

1. Poor presentation
2. Difficult political circumstances
3. Inappropriate timing
4. Inadequate budget
5. Lack of available resources
6. Competing priorities
7. Aversion to risk

Yet it is important for an organisation to encourage suggestions for new projects. Conceptualisation of new ideas supports innovation and entrepreneurship and enables an organisation to keep moving forward. Many believe that the importance of translating ideas into innovations has never been greater (Ahn & Meeks 2007; Angra, Sehgal & Noori, et al. 2008; McKinsey 2004).

Formal approval of the project concept and announcement of the decision to develop a business case could take the form of something as simple as an email, an item in meeting minutes or even a statement in a public address. The funder may well qualify approval of the project concept by stressing the importance of the current problem or the opportunity to be explored. It would be expected that the eventual project objective would reflect such a comment. The approval should also clearly identify the project champion, and confirm that he/she is in charge of business case development.

5.1.4 The Role of the Business Case

A business case is a formal way of asking "Because we seek a particular set of outcomes, we have to produce and utilise a certain collection of outputs. Producing those outputs will require significant work and resources. Are you willing to fund the exercise?" In other words, when specifying a project, *target outcomes determine the outputs that are to be produced.* Paradoxically, the original idea for a project usually arises from a contrasting sort of statement, along the lines of "If we had available to us a certain output, then what desirable end effects might we be able to realise?". In other words, when conceptualising a project *outputs may suggest desirable outcomes.* Specifying a project, however, involves a subtle, progressive shift away from a *desirableoutcomes might follow outputs* mentality towards a *scope depends on targetoutcomes* analysis. It does not really matter if outputs drive the early thinking, as long as the eventual business case is driven by target outcomes.

For this purpose, the seven project elements defined in Chap. 4 should be addressed and included (together with other critical supplementary details) in the business case:

1. *Scope.* A project's scope is set when its target outcomes and committed outputs are all identified and validated. More will be said about validation of a project's scope in the discussion of the utilisation map below.
2. *Stakeholders.* Project stakeholders are identified, their respective stakeholdings analysed and an engagement plan assembled for each.
3. *Risks.* Threats are identified, analysed and, where appropriate, made the subject of a risk mitigation programme.
4. *Issues.* Issues are identified and a decision taken on how each is to be resolved.
5. *Schedules.* Estimates are provided of the times required to deliver the project's outputs and to secure its target outcomes. Both these are supported with a schedule of milestones.
6. *Resources.* An estimate is made of the resources required in three areas: to create outputs, to facilitate their utilisation and to operate any that will be introduced into ongoing business operations.
7. *Governance.* A project governance model is proposed for the management and leadership of all those who have a part to play in the project.

These elements should be covered in detail in the business case. For example, the scope element is handled using a list of outputs that forms part of, what is called *the statement of scope.* Because each output must be completely specified before it is (eventually) produced, a foundation is laid in the business case for such a specification. This is achieved by listing the critical fitness-for-purpose features of each output. (The concept of fitness-for-purpose is discussed in Sect. 5.2). In most cases, the detailed specifications for all outputs are prepared much later (as part of the execution phase).

The work involved in developing each of the seven elements listed above tends to be heavily influenced by the work being done on the others. As more is discovered about one element, further information is inevitably uncovered about others that have already been considered in earlier activity. For example, when deciding how to engage stakeholders, we might find it necessary to produce outputs that had not previously been regarded as in-scope, (such as programmes of professional development for staff). In that case, stakeholder analysis results in additional outputs for the project's scope. Because project elements are so heavily interconnected, the process of developing the business case should allow for *iteration.*

5.1.5 Developing the Business Case

The assembly of the business case can be done by adapting the spiral approach to development (Boehm, 1988). This offers a systematic way of progressively building on information already obtained, until everything required for a funding decision has been assembled. The spiral approach to project initiation suggests that material for the business case is best prepared in a large number of short working sessions, rather than a small number of long sessions. Anecdotal evidence suggests that the teams involved in this process usually find an iterative approach more rewarding, interesting and satisfying than lengthy sessions conducted for the sole purpose of "filling out the templates". The end product from the latter approach is often an incomplete, unreliable and irrelevant document that is quickly pushed to one side as having no bearing on what the project is really about.

Figure 5.2 shows the general structure of the spiral approach to project initiation. Seven of the arms are associated with project elements, while the eighth is concerned with integrating these elements into a business case. The spiral begins at the centre of the diagram and progresses outwards clockwise. Each iteration (cycle) adds information along each arm. The further along each arm the process progresses, the greater the amount of detail being assembled for the associated element. Each point of intersection between the spiral and an arm represents a working session for the initiation team. It is desirable to make each of these working sessions short. It should be noted that the spiral continues after acceptance of the business case, into project planning. The use of the spiral beyond initiation is covered in Chap. 6.

As a matter of expediency, the first element to be addressed as soon as project definition begins is issues. This is because issues start arising from the moment work begins on the embryonic idea for a project, even before any working papers are produced. The last element to be addressed in each cycle involves a packaging step, assembling all elements into a business case.

In practice, it is not necessary to adhere strictly to the order in which the spiral intersects the arms. For example, on iteration #2, having completed work on stakeholders, it may be appropriate in some cases to go directly to governance, rather than deal with risk next. Each working session (represented by dots in Fig. 5.2) involves a peculiar collection of activities that depends on two factors: the arm on which the dot falls (which determines the topic of the session) and the distance of the dot from the origin (which determines the amount of detail to be covered).

Successive orbits of the initiation spiral involve additional working sessions and result in further development of all the core components of the business case. It is common practice to use templates for recording the details about each element of a business case. Templates provide a check-list of pre-determined items of information that are required (as well as suggesting a layout for that material). For purposes of explaining the scope of each iteration in the spiral, it should be noted that, in general, the templates associated with each component of a business case fall into two classes: those that have a tabular format, and those that do not. Templates of the first type are called "registers", presented as a table in which columns are associated with attributes and rows with instances. For example, a typical stakeholder register has columns for name, stakeholding, engagement plan and so on (refer to Sect. 4.5 in Chap. 4 for a discussion of stakeholder analysis and

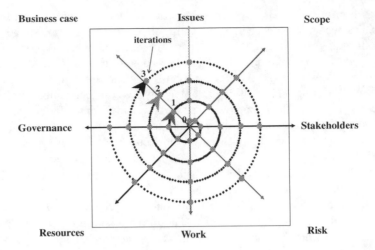

Fig. 5.2 The spiral approach of project initiation

engagement), while each row identifies a particular stakeholder. By way of contrast, the template for a statement of scope (Sect. 5.2) is not of this form (because at any one time there is only ever a single instance).

Registers relate in a very natural way to the iterative nature of the initiation spiral. For example, Fig. 5.3 shows how the information contained in a typical stakeholder register might be developed progressively with each iteration of the spiral. In this case, the first working session is confined to the identification (by name) of those entities who are candidate stakeholders in the project. In the second, these entities are analysed to obtain information about their stakeholding, generic roles in the project and any issues that need to be noted. Later, in the next iteration, after information has been gathered from other arms of the spiral, a stakeholder engagement plan is developed. In each cycle, extra detail for the business case is assembled by completing more of the columns in the stakeholder register, to the right of those already considered. With each iteration, not only is the range of columns (attributes) expanded, but the number of rows is also increased, as more potential stakeholders are discovered. It is also possible that some of those who were identified in earlier iterations as stakeholders are now found not to be, and so the associated rows would then be deleted. A question now arises: "Which columns in a project register are handled in which iteration of the spiral?" This is decided on a project-by-project basis, but clearly, there are two extremes. All columns can be considered in one (the first) iteration or one column can be considered with each iteration. (In the latter case there will at least as many iterations as there are columns). In general, work on a column should be delayed for a later iteration when the collection of instances (rows) assembled to date is expected to undergo significant change (through deletions and additions). This principle will reduce the amount of unnecessary work carried out on a register. For example, there is little point in spending time on assembling an engagement

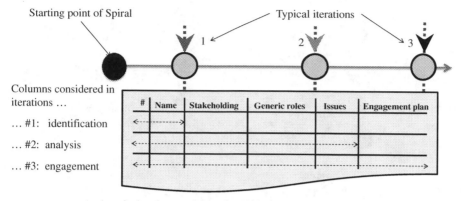

Starting point of Spiral

Typical iterations

1 2 3

Columns considered in iterations …

… #1: identification

… #2: analysis

… #3: engagement

#	Name	Stakeholding	Generic roles	Issues	Engagement plan

A typical register-style template (this one concerning stakeholders) in which:
- rows are associated with particular stakeholders
- columns with their attributes.

Fig. 5.3 The business case development process—using the example of a stakeholder register

programme for stakeholders if the team suspects that many of those in the current list will be culled as the scope of the project becomes clearer.

Not all components of a business case are supported with a register (the statement of scope is a case in point because it does not have a tabular structure of instances and attributes). Despite this, such components are also the subject of an iterative process during initiation, but, in those cases, each iteration results in the element being simply "refined" (rather than "expanded").

5.1.6 Expectations, Constraints and Assumptions

An important role of executives during project initiation is to set appropriate expectations from a project. However, expectations must not be confused with project constraints. For example, Charles Edwards (the CEO of ICO Inc. in the BuyRite case study) may wish to impress the Board with an ambitious target date for "fixing" the procurement problem. To lay the ground work for such a commitment, he may be tempted to set an arbitrary date as a mandatory "constraint" in his approval of the early Project BuyRite concept. It would be more appropriate to state this as a managerial expectation. By way of contrast, consider an exercise to rebuild an existing suburban train station where construction activity can only take place from 23:30 h through 04:30 h overnight (when trains are not running). In this case, there is an absolute constraint of 5:00 h/day on site access. Framing expectations as constraints frequently results in the quality of business cases being compromised and projects being exposed to unacceptable risk.

Expectations relate to the perceptions and beliefs of stakeholders about the project. Expectations become of concern to key players when they deviate

Box 5.2 A Case Study: The Queen Mary 2

The Impact of Arbitrary Constraints on a Project

The Queen Mary 2 (the largest ocean liner ever built) provides an example of a project that was initially infeasible because of constraints. In the early design brief for the ship, constraints were set for: width (to pass through the Panama Canal), height (to fit under the Verrazano-Narrows Bridge in New York City), fuel efficiency, stern profile (similar to that of the Queen Elizabeth 2), minimum number of passengers, number of cabins with balconies and speed (Barron 2004).

Extensive analysis by the naval architect eventually revealed that there appeared to be no feasible design option. As a result, the funder (Carnival corporation's CEO) relaxed the constraint related to the ship's width, thus making the project technically feasible (Barron 2004). The removal of the width constraint, however, meant that the trip time from Europe to the US West Coast was increased (because of the need to travel around Cape Horn). The impact of this change on the project would have been to reduce somewhat the size of the market that the Company could access with its new service.

It should be noted in passing that the scope of this project would have gone well beyond the ship itself, to encompass such outputs as: operational processes, maintenance facilities, promotional programmes, crewing arrangements and a product development business unit. The initiative is usually identified as "The Queen Mary 2 Project" because the ship clearly dominated the entire exercise.

significantly from the business case, because unmet expectations can themselves create undesirable outcomes (such as reduced public support for the funding organisation). Managing expectations, widely acknowledged as an important facet of project management, can be viewed simply as a process of ensuring that all stakeholders understand relevant parts of the business case.

Managing expectations is not to be confused with gaining support. Consider a project in which a particular stakeholder "S" is implacably opposed to the initiative because he wants to have output "D" included, but, at the same time, fully expects the scope of the project not to include "D". In this situation there might well be a stakeholder management issue (because "S" does not agree with the business case), but there is no expectations management issue (because the expectations of "S" about the project align completely with the business case).

A common problem arises when arbitrary (and frequently unrealistic) constraints are set by funders and other key stakeholders early in the project.

Too many constraints can reduce project feasibility or its potential worth, as is demonstrated in Box 5.2 about construction of the Queen Mary 2.

Assumptions relate to conditions that are accepted as "givens" in the business case. For example, a project involving the importation of engineering equipment may be approved on an assumed exchange rate. If an assumption is later found not to be true, then changes have to be made in the business case (with a corresponding change in the attractiveness of the project).

5.2 Scoping the Project

Scope is the foundation of the business case, because it establishes a project's boundaries and hence directly influences outcomes and outputs, as well as duration, budget and level of risk. *Scope creep* is claimed to be common cause of project failure. Scope creep takes the form of a gradual expansion of a project (sometimes without any corresponding addition of target outcomes, increase in budget or extension to duration). Unless there is a clear definition of a project's scope, this phenomenon can be not only difficult to detect, but even more difficult to manage. We propose that *a list of outputs* offers the only reliable foundation for scoping a project, however, before a project's outputs can be finalised, the outcomes to which they contribute must be established (using the ITO model described below). With a set of agreed target outcomes, the proposed scope of a project can be validated. That observation does not, however, imply that outcomes must be set *before* outputs can be considered. As pointed out earlier, the embryonic ideas that underpin projects are often expressed in terms of outputs with the result that discussion of outcomes happens later. The initiation spiral described above allows for that type of iteration.

5.2.1 Setting Project Scope

If stated appropriately, a project's scope enables one to decide, in effect, "where the boundaries of the project lie". While it is relatively easy to assemble a statement of scope for a project, it is much more difficult to confirm that statement is reliable. One of the first questions to arise during initiation is: "Given all of the alternative lists of outputs that might be proposed for a project, which (if any) are correct?" This question is widely recognised as critical to the way a project unfolds, and to its eventual success, but the existing literature is generally silent on how it is to be answered. The process we propose to resolve this issue has two parts: develop a statement of scope and validate it, along the lines of that shown diagrammatically in Fig. 5.4.

In its early versions, the statement of scope includes (amongst other things) a simple list of candidate outputs. Later this will be supplemented with supporting lists of fitness-for-purpose features (one such list for each output).

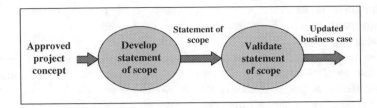

Fig. 5.4 Project scoping processes

Various stakeholders often make irreconcilable demands on a project by proposing their own list of outputs (or lists of fitness-for-purpose features), driven by explicit or implicit objectives and associated desired outcomes. When the early views of a project give rise to conflicts of this nature, it is necessary to decide which declaration of scope (if any) is correct because an "incorrect" scoping statement will lead to one of two problems:

- *Overscoping.* Whereby the current scope includes redundant outputs (or outputs with superfluous features). Overscoping is commonly called "over-engineering".
- *Underscoping.* Whereby the current outputs cannot support target outcomes (either because some are missing, or because some do not have all the characteristics that allow project customers to utilise them effectively).

Project scope is established using one tool (the statement of scope) and validated with another (the utilisation map):

1. *The Statement of Scope.* The statement of scope declares the project's objective, and supports this with lists of target outcomes and committed outputs.
2. *The Utilisation Map.* The utilisation map validates the statement of scope by confirming that all of the proposed outputs do contribute to the generation of target outcomes and that all target outcomes can be generated with the proposed project outputs. The statement of scope and utilisation map are assembled during initiation and eventually included in the business case.

We now explore the scoping process by discussing in turn:

- The statement of scope and its major components (Sect. 5.2.2)
- The identification and definition of target outcomes (Sect. 5.2.3)
- The identification of committed outputs (Sect. 5.2.4)
- Scope validation (Sect. 5.2.5)
- The definition of committed outputs (Sect. 5.2.6)

5.2.2 The Statement of Scope

Before discussing the scoping and validation process, we need to consider a structure for a scoping statement.

While there is general agreement in the existing literature that a scoping statement should take the form of a list, there is a somewhat wider range of views about what the list should contain. Suggestions include objectives, deliverables, resources and justification for the projects (Kerzner, 2009).

Because a list of activities can be derived (via a WBS) from a list of deliverables (outputs), and because, in turn, a list of resources can be derived from a WBS, it appears that a list of outputs provides the only reliable foundation for a statement of scope. Accordingly, we propose a scoping principle whereby:

1. A project is scoped if and only if its outputs are defined.
2. Project outputs are defined when two conditions are met:

   – all outputs are listed
   – critical fitness-for-purpose features are set for each.

The statement of scope is a formal way of declaring "what is in and what is out of the project". For such a declaration to be effective it should have three components:

- An *objective statement* that establishes succinctly why the project is being proposed.
- A *list of target outcomes* that, if realised, would imply that the objective had been achieved.
- A *list of outputs* that are to be delivered.

The first of these three is discussed here, and the other two below. We propose a format for the objective statement based on the following principles:

1. *Confirms project rationale.* Provides an overarching answer to the question "Why is this project being funded?" and, accordingly, begins with the word "To ..."
2. *Simplicity.* Takes the form of a slogan or "one-liner". This makes it suitable for incorporation into project documents and publications, such as newsletters. In this form, it is also easily memorised by all stakeholders.
3. *Captures project intent.* Expressed in outcome-terms—consider a project for which the objective statement is: "The objective of the ABC project is to install the XYX computer system". Because it is stated in output (not outcome) terms, it would fail this last criterion. By way of contrast, "The objective of the EFG project is to reduce stockouts and working capital" would satisfy this criterion.
4. *Aligned with the organisation's strategic goals.* For example, the balanced scorecard (Kaplan and Norton 2005) can be used to define project objectives that are closely aligned with the organisation's vision and goals.

Although an objective statement would appear as the first of the three components in a typical template for a statement of scope, it is best tackled last (because it is, in effect, a summary of the project's target outcomes).

The statement of scope is the first substantive piece of project documentation prepared during project initiation. Early versions of a statement of scope are

Preliminary Statement of Scope - The Project BuyRite Case study
1. Project objective:
– To transform ICO's procurement unit into a world class operation.
2. Target outcomes:
Reduced procurement costs
– Reduced payment times to our suppliers
3. Committed outputs:
– New procurement processes
– A panel of preferred suppliers
– A new office for the Procurement Department

Fig. 5.5 An initial (tentative) statement of scope for project BuyRite

typically not only incomplete and inconsistent, but also relatively unreliable because they are assembled when information about the project is fragmented and sketchy. Figure 5.5 presents an initial (and highly tentative) statement of scope for Project BuyRite. It is expected that, before this is finalised, new items (particularly outputs) may be added, while some existing items may be deleted. As will be seen later, when this version of the initial scoping statement for Project BuyRite is later validated, it is found to be incorrect, requiring the addition of new outputs and the deletion of others.

5.2.3 Identifying and Defining Target Outcomes

Early working versions of the statement of scope are usually confined to lists (of outputs and target outcomes), because the first task is to simply *identify* relevant items. Later versions however, will be supported with definitions of those outcomes as well. Outputs are defined by listing their fitness-for-purpose features, while outcomes are defined by setting a number of core attributes. Both these topics are discussed later. As a rule of thumb, the number of target outcomes for a project should not exceed five. This guideline is offered because additional outcomes:

- Require an increase in scope.
- "Dilute" the attention of key stakeholders.
- Increase the likelihood that the business case will not be realised (and hence reduce its acceptability).

A target outcome is a desirable end effect that is actively sought by a funder and so Project BuyRite, for example, has target outcomes that relate to ICO's desire to reduce operating costs and improve its image in the industry. Target outcomes are identified very early in initiation but defined towards the end of that phase.

The process of identifying target outcomes involves a number of steps. The very early exploratory work carried out in conceptualisation purposely seeks the views of a wide range of potential stakeholders, resulting in, what is often, an unworkable large number of candidate target outcomes. (Certainly, the list will grow very quickly to exceed the suggested limit of five). At some point, it becomes necessary then to cull this list, so that only a small number have to be defined in the eventual business case. It is suggested that this culling not be done too early, to encourage wide discussion of the project's eventual shape. Four criteria are suggested to act as filters on the working list of candidates:

1. *Importance*. An outcome has to be important to the funder if it is worth targeting.
2. *Measurability*. If an outcome is difficult or costly to measure, or if the proposed measures are difficult to interpret, then it does not meet this criterion.
3. *Lag*. If an outcome can be detected only after an "unworkably long" period of time, then it is not suited to targeting. (It is up to the key players to decide what represents an acceptable lag).
4. *Plausibility*. If the causal linkage between the project's outputs and the candidate outcome is "weak", then it fails the plausibility test. (This is as close as we get in this methodology to a "test of causality").

Because of their importance, target outcomes should be carefully defined by setting seven attributes for each, as shown in Table 5.2, which illustrates the concept by defining an outcome from Project BuyRite.

Target outcomes involve two points of measurement: (one relating to a scenario where the project does not proceed and another relating to the scenario where it does proceed) and so they are usually expressed as "increased ..." or "decreased ...". Thus in Project BuyRite a target outcome is "*Reduced* procurement cost". There are some exceptions to this convention, especially when a target outcome is binary. Take a project triggered by legislation that requires a company to develop and maintain a register of all political donations. An appropriate title for the target outcome of such an exercise would be "*Compliance with* the political donations legislation".

The word "improved" should be avoided in the title of target outcomes because it is unnecessarily vague. (Projects seek changes in the measures adopted for certain variables and a change in a measure implies that it either increases or decreases). "Improve" is, however, useful and in some cases appropriate, for objective statements and in the names of certain outputs. For example in Project BuyRite the core output could well be described as an "improved procurement process".

In most projects it is also important to have corresponding measures taken from the *current* scenario. A current measurement of a variable used to define a target outcome is called a baseline, and the process of making the measurement is called "baselining". Here the term *baseline* (as a process of measurement) is to be distinguished from the concept of a *baseline document* (where a project is specified). Without a baseline, the nature/magnitude of the current situation cannot be

Table 5.2 The attributes required to define a target outcome

Attribute	Meaning	Example for Project BuyRite
Title	Usually this will begin with "reduced" or "increased", however there are some other acceptable forms (such as "compliance with …")	Reduced procurement costs
Description	Some relevant detail on the target outcome on the title. There may also be further detail about how this outcome is to be measured	The average cost of filling a single purchase order
Measure	The units or dimensions that will be applied	$ per purchase order
Target	This will become a threshold in the business case. It can be expressed in absolute (fixed value) or relative (compared to another value) terms	A relative target for Project BuyRite a target might be a reduced cost per order of 25%
Source/method	Outline of how the data to measure the target outcome will be acquired or where it will come from	Two of ICO's information systems: "cost accounting system" and the "procurement system"
Achievement date	This will become a threshold in the business case, indicating the latest date by which the target outcome is to be secured	One year from project approval
Person accountable for realising the target outcome	Usually the project owner (who is appointed by the project funder)	Nancy Palmer, National Procurement Manager

determined empirically, and so the rationale for the project may be difficult to establish.

Measurement itself involves time, resources and possibly outlays, which must be included in the estimates appearing in the business case. This raises an important issue: When/how are baseline measurements to be made? Consider the situation that ICO would face in Project BuyRite if the current and predicted costs for procurement were not known. Because the target outcome is expressed relatively (-25%), how does Charles Edwards (the funder) know if that is worth achieving? 25% off a low base could imply low benefits and a project of low value. There are two options:

1. Assume that the reduction would be worthwhile and accept the business case subject to an early phase of baseline work confirming the correctness of the assumption. This approach would have to acknowledge that the baselining may reveal a project of low worth, in which case the exercise would have to be abandoned at that point.

Box 5.3 Practicalities

Customers should Participate in the Discussion of Project Scope, but They Cannot Define it

Project customers (those who generate target outcomes through their util-isation of the project's outputs) are clearly critical stakeholders in a project. If they do not utilise the projects outputs, then the project cannot generate its target outcomes. A long-held, and often repeated, view of project customers leads to the conclusion that they therefore have a dominant role in defining the project's scope and specifying its outputs. Such an assumption sits behind a number of popular (but rather naive) approaches in which "end users scope the project".

The coverage of initiation here notes that scope has two parts: listing outputs and deciding in their key features. Any methodology, which seeks to scope a project by asking customers to identify outputs, is invalid because it involves circular reasoning. The ITO model shows that customers become known only *after* outputs are identified, not *before*. In other words, outputs identify project customers, not the other way around.

Whoever is commissioned to scope a project must: (1) have a clear view of the funder's objectives, (2) face no conflict of interest with generation of those outcomes and (3) thoroughly understand the mechanisms by which outputs and outcomes are related.

So where does this leave customers as far as scoping a project is con-cerned? It does not mean that they are irrelevant, quite the opposite. Cus-tomers should be involved in scoping, but only to identify certain fitness-for-purpose features that proposed outputs must have so that they can be utilised successfully. In other words, customers have an important role in *con-straining* the project's scope, but a limited role in *setting* scope. In general, any congruence of views about the project between the funder and a cus-tomer is fortuitous. In many cases customers are negatively impacted by a project or will have objectives that conflict with those of the funder and so considerable care must be exercised when accommodating their views about desirable and undesirable features of outputs.

2. Approve baselining as part of the work on the business case. This means a longer and more expensive definition phase. It also raises the prospect of project definition being abandoned if baselining reveals an unattractive investment.

In some cases, especially when target outcomes have been identified by dif-ferent stakeholders, it may be found that some of them are negatively correlated. Consider a project to improve service levels to an organisation's customer and, at the same time, reduce operating cost. Because, in general, these two variables are inversely related, when deciding on target levels, a trade-off must be made between them. In such circumstances, it may even prove necessary to abandon one of them.

5.2.3.1 A typology of target outcomes

Our experience with development of target outcomes over a large number of projects suggests that there is an underlying typology. The exemplars in Table 5.3 represent some commonly occurring outcomes, arranged here as a (suggested) hierarchy.

5.2.4 Identifying Committed Outputs

The scoping principle states that outputs are defined when two conditions are met: they are listed and their fitness-for-purpose features are specified. The first condition is met by identifying the outputs that are to be produced, while the second is met by specifying the fitness-for-purpose features of each output in the list.

Figure 5.6 shows the process of scoping a project's outputs.

In this section we focus our attention on the "identify outputs" process, leaving the "define outputs" process until Sect. 5.2.6.

According to the ITO model, target outcomes are generated when outputs are, in due course, utilised by the project's customers. This seems to suggest that a project's outputs can be identified only after target outcomes are clearly identified. Such an inference is true for some outputs, but there may well be others that are required for reasons that have little to do with the generation of target outcomes.

In Table 5.4, we propose a typology of all the outputs that can be created in the course of a project, involving just seven distinct classes, only five of which may appear in a statement of scope. Unlike the typology of outcomes (which appears to be open), this one appears to be closed. In other words, we believe that it is exhaustive.

5.2.5 Validating Project Scope

Validating a project's scope is the process by which the "correctness" of a scoping statement is judged. The scoping principle discussed above says nothing about whether or not a project's scope is right, thus a statement of scope can be set for a project which, although clear and unambiguous is also "incorrect", and so a discussion is required about the concept of the "correctness" of a scoping statement. This is then supported with the outline of a process for validating a proposed scoping statement.

A statement of scope for a project is correct if four conditions are filled:

1. No outputs are redundant.
2. No outputs are missing.
3. No output has superfluous fitness-for-purpose features.
4. No output is missing any necessary fitness-for-purpose features.

The process of validating a project's scope considers both ITO and non-ITO outputs separately:

A. To validate ITO outputs: Apply the utilisation map to identify redundant and missing outputs (discussed below).
B. To validate non-ITO outputs: Scan for candidates by considering, in turn: the risk register, the stakeholder register, the policy/regulatory environment surrounding the project and dependencies amongst the outputs.

5.2.5.1 The Utilisation Map

We now discuss in detail the validation process for ITO outputs. To understand the utilisation map it is necessary to revisit the ITO model. A project is funded to realise a flow of target outcomes, and outcomes are generated when customers utilise the projects outputs. If we can model the relationships amongst outcomes, outputs and customers in an appropriate way, then redundant and missing outputs may be revealed.

The utilisation map takes the form of a table in which:

1. Each column is associated with a target outcome.
2. Each row is associated with an output.
3. Each cell identifies (as a list) all those project customers who utilise the output on the left to generate the outcome at the top.

In the following discussion the term "customer" is qualified according to whether we are discussing the "project customer" (one who utilises one or more project outputs) or the "organisation's customer" (an entity or organisation that pays for services or products provided by the organisation).

The cells in a utilisation map can finish up with any combination of three sorts of entry:

1. *Blank (the cell is empty).* No one utilises the output at the left to generate the outcome at the top. For example, in the case study introduced in Box 5.4, no one utilises the panel of preferred suppliers to reduce payment times to suppliers.
2. *A single utilisation.* One project customer utilises the output to generate an outcome. For example, only procurement staff utilise the programmes of professional development to reduce costs of procurement.

Table 5.3 A proposed typology of project outcomes

Class	Subclass	Examples of outcomes
Business outcomes	Sales & marketing	Increased sales volumes
		Reduced costs of goods sold
		Increased market share
		Reduced rates of account attrition
		Increased levels of customer satisfaction
	Operations	Increased product/service quality
		Reduced rates of product failure
		Reduced lost-time injuries
		Reduced rates of staff absenteeism
	Client access (to products/services)	Reduced service fulfilment time
		Reduced costs (to client) of service fulfilment
		Reduced effort required of a client for service fulfilment
	Preparing for the future	Increased number of potential markets where the company can sell its products
		Increased number of products/services the organisation offers
		Increased understanding of new technologies
		Reduced time-to-market for new products
Social outcomes	Crime	Reduced incidence of violent crime
		Reduced rates of recidivism
	Health	Reduced rates of child mortality
		Reduced numbers of smokers
		Increased awareness of dangers of drug abuse
	Sustainability	Increased use of reusable water
		Reduced volumes of rubbish going to landfill
	Education	Increased proportion of students completing high school
		Increased numbers of teachers meeting agreed competency standards

Table 5.3 (continued)

Class	Subclass	Examples of outcomes
National outcomes	Security	Reduced risk exposure to acts of terrorism
		Reduced response times to international requests for security assistance
	International standing	Increased numbers of overseas visitors
		Increased levels of participation in regional affairs
	Environmental	Reduced levels of infestation by exotic plants
		Increased stocks of threatened native species

Table 5.4 A proposed typology of project outputs

Major class	Minor class	Output	Definition/example
Deliverables			Outputs that may appear in a statement of scope. These are also called final outputs. Final outputs are necessary for project success, (but perhaps somewhat paradoxically, these are the only outputs recognised when judging project *management* success)
	ITO outputs		Required by the ITO model for the generation of target outcomes
			In Project BuyRite the new procurement process is an ITO output
	Non- ITO outputs		Required of the project, but are not utilised to generate target outcomes
		Risk mitigation outputs	Required as part of the project's risk mitigation programme
			There is concern that the proposed reduction in the number of suppliers might weaken ICO's future negotiating position. To address such a risk, it is decided that at least two suppliers of strategically critical products/services must be appointed to the panel (representing an additional fitness-for-purpose feature)
		Stakeholder engagement outputs	Required as part of the project's stakeholder engagement programme
			It has become clear that the new BuyRite arrangements will require a smaller Procurement unit and so a number of existing staff should be reassigned, (or offered redundancy). Accordingly, a number of redeployment/redundancy packages are to be assembled for staff who will not remain (an extra output related to the engagement of those staff in the project)
		Mandatory outputs	Required by law, policy or regulation
			Internal audit has uncovered some weaknesses in the way that procurement staff record details of their contacts with suppliers and they have recommended that a more formal approach be adopted. To comply with this recommendation, a new set of "vendor contact" procedures will be produced (an additional output)
		Dependent outputs:	Required by another output or project
			The project requires some specialised systems infrastructure and a number of application systems that run on that infrastructure. While it may appear that no-one uses the technical equipment (or at least not directly), it is clear however that such infrastructure is required by another output (the application system) and so it would make more sense to "bundle up" the infrastructure and software into a single "information technology" output

Table 5.4 (continued)

Major class	Minor class	Output	Definition/example
Procedural outputs			Outputs that do not appear in a statement of scope. In these cases, the output is of no interest after the execution phase of the project is completed. In other words, project management success does not involve consideration of procedural outputs
	Intermediate outputs:	Required for the production of final outputs, but have no ongoing role beyond that	A specification for a set of tests that will be used for a new process is an example of an intermediate output
	Above-the-line outputs	Required as part of the overarching planning and management framework	A Gantt chart is an example of an above-the-line output

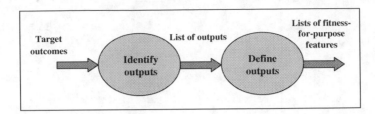

Fig. 5.6 Output scoping processes

3. *Multiple utilisations.* Several project customers utilise the output to generate an outcome. For example, reduced payment times to our suppliers require that the new procurement processes be utilised by both the procurement staff and suppliers.

During validation, the utilisation map is employed in the following way:

1. For each outcome column, check to see if the combined contributions of all utilisations listed anywhere under this outcome represent an acceptable level of achievement. The extreme case is where a column is empty (that is, no project customer is identified as utilising any output to generate this outcome). This could indicate missing outputs (that is, the project is currently underscoped). Before expanding scope to correct this problem a number of other explanations should be considered:

   – Some project customers have been overlooked. This can happen either because they have not been identified or because their contribution to an outcome has been underestimated. (In either case, new entries are required into the relevant cells).
   – The outcome is not really sought by the funder (unlikely, but the question may need to be asked).

   If the addition of extra outputs proves necessary then it is clear that, relative to the current scope, the resulting expansion of the project will increase cost and duration, lowering, to some extent, the project's worth and attractiveness as an opportunity for investment.

2. For each output row confirm that the combined contributions of all utilisations listed anywhere against this output represent a significant role for it. The most extreme case is where the row of the utilisation map is empty (that is no project customer is identified as utilising this output at all). This could indicate that the output is redundant (that is, the project is currently overscoped). In view of this, before removing the output from the current scoping statement to correct the situation, a number of other explanations need to be considered:

   – Some project customers have been overlooked. This is an identical situation to that discussed for the "empty column" situation above.
   – A non-ITO output is involved (and is therefore unrelated to a target outcome).

Box 5.4 Illustration of Concept: A Preliminary Utilisation Map

Project BuyRite Case Study

Based on the tentative statement of scope provided in Fig. 5.5, we can construct the following (equally tentative) utilisation map as Table 5.5:

Table 5.5 Tentative utilisation map for Project BuyRite

ITO output	Target outcome	
	Reduced procurement costs	Reduced payment times to our suppliers
New procurement processes	Procurement department staff Suppliers	Procurement department staff
A panel of preferred suppliers	Procurement department staff	N/A
A new office for the Procurement Department	N/A	N/A

Three aspects of this utilisation map are worth noting. The first is that all three types of entry discussed above are represented here. In particular:

- The bottom left entry is blank because a judgement has been made that no-one generates the target outcome of reduced costs by utilising a new office.
- The entry in the second row/first column confirms that only the procurement staff utilise the panel of preferred suppliers to reduce costs.
- The top left entry indicates that staff from both ICO and suppliers will generate reduced costs (for ICO) by utilising the new procurement process.

The second observation is that the two target outcomes appearing in this particular utilisation map are inversely related. Inverse outcomes are those where a desirable change in one can be gained by consciously sacrificing the other. For example, supplier payment times can be reduced with extra staff in the Payables Department (thus increasing procurement costs).

The third observation is that because subsequent analysis reveals this initial view of Project BuyRite to be incomplete—the version of the Utilisation map appearing in the illustrative business case below has a different (more reliable) structure than the version shown here.

Table 5.6 The final statement of scope for Project BuyRite

Statement of Scope—The Project BuyRite Case study

1. Objective statement:
   To achieve world class performance in our procurement operations
2. Target outcomes:
   Reduced procurement costs
   Reduced payment times to our suppliers
3. Committed outputs:
   New procurement processes
   New procurement policy and procedures manual
   Enabling applications systems and new technical infrastructure
   A panel of preferred suppliers
   A restructured procurement unit
   Programs of accredited professional development for staff

- A target outcome is missing. This can happen when a clear role for the output is understood, but not explicitly covered with a defined target outcome.

When the utilisation map passes all these "tests" it can then be used to finalise the project's Statement of Scope and to include it as part of the business case.

Table 5.5 suggests a utilisation map for Project BuyRite where the following points should be noted:

- The two outcomes are negatively correlated, in that, reduced costs of procurement could be realised through strategies that would cause increased payment times to our suppliers, and reduced payment times to our suppliers could be realised through strategies that would cause increased costs of procurement. Such a situation demands that targets be set for each that are not mutually exclusive.
- A blank cell links the output "A panel of preferred suppliers" to the outcome "Reduced payment times to our suppliers". This indicates that, while it cannot contribute to the outcome, the output remains in scope because of its contribution to the other outcome "reduced costs of procurement"
- The last output in the table ("New office for the procurement department") is shown as not connected to either outcome and so is a candidate for removal from project scope. The issue here is not so much that there is no causal link, but that any link is weak. That determination is peculiar to the project's circumstances and is a judgement that must be made by those doing the analysis.

A Statement of Scope for Project BuyRite is suggested in Table 5.6.

According to the scoping principle, the scope of a project requires not only a list of outputs, but also preliminary specification for each, in the form of a list of fitness-for-purpose features. Just as a "wrong" list of outputs can give rise to underscoping or overscoping, similar situations can arise with fitness-for-purpose lists. To address this issue we need to recall that a project's outputs can be categorised into two: ITO outputs (required by the ITO model for the generation of target out-comes) and non-ITO outputs (required by the project for other reasons, such as risk mitigation). The fitness-for-purpose features of non-ITO outputs are decided simply by asking, "what characteristics does this output need so that it can fill its intended role?". For ITO outputs, however, we propose a tool to help address this issue, called the "utilisation storyboard" which is discussed below.

5.2.6 Defining Outputs

Following validation of outputs, in its final form, a statement of scope goes beyond the identification of each of the project's committed outputs, it also provides the nucleus of a specification for each output, in the form of a list of critical fitness-for-purpose features. A fitness-for-purpose feature is a characteristic that the output must have so that it can be utilised by project customers in such a way that target outcomes are generated. Consider, for example the project described in Box 5.5.

Box 5.5 A Case Study: Historic Australian Goldmining Township

The Concept of Fitness-for-Purpose

An Australian state tourism authority has decided to promote a historic goldmining township to middle-class residents of Shanghai. During the mid-1800s, Chinese flocked to the newly-discovered goldfields to work as prospectors and miners. Many of these came from Shanghai, creating important cultural links between the two places.

The township has been classified by the National Trust and so there are restrictions on what can be done. For example, signage must "harmonise" with the existing architecture. Residents have expressed wide-ranging views about the desirability of the exercise.

The project will also involve heavy promotion in Shanghai through TV, radio, newspaper and the internet.

Target outcomes from the venture include: increased awareness of Australia as a tourist destination (amongst Shanghaiese) and increased visitation to the township. ("Visitation" is a defined term in tourism industry parlance meaning the overall numbers of visitors to a target location). Outputs include: a "heritage trail" of distinctive signs, plaques at points of interest, multimedia booths (serving as interpretive stations) and souvenir stalls. The eventual business case for this project will list critical fitness-for-purpose features for each of these outputs.

Even superficial analysis quickly reveals that having a Chinese translation is a fitness-for-purpose feature of the plaques used to identify points of interest. This now raises an important question: "How are reliable lists of fitness-for-purpose features to be assembled?". While the derivation of appropriate fitness-for-purpose features for an output is a largely creative process, it can be supported with some analysis. We propose one such tool, the utilisation storyboard, and describe below how it might be applied to this particular case study.

5.2.6.1 The Utilisation Storyboard

The utilisation storyboard is based on the utilisation map. It provides a descriptive outline of the way that each output is utilised by a project customer. It is best introduced with an illustration, which we then follow with a general discussion.

Consider the tourist case study introduced above. It should be clear that visitors from Shanghai are important project customers. (They also happen to be "customers" in a more conventional sense, but that is not always true). It is also to be expected that a utilisation map (not shown explicitly here) would reveal that these project customers utilise all four of the outputs to generate both outcomes identified in the description of the case. Take the outcome "increased visitation to the township" and the output "plaques at points of interest". Here, the utilisation storyboard might appear as follows (Table 5.7).

A storyboard is required for every entry in the utilisation matrix, that is, for every appearance of a project customer in any cell. In this example, the storyboard links the Shanghai tourist to the plaques and the visitation outcome. Because the utilisation map may have many (repeated) entries, in general, each output, outcome and project customer will all appear in multiple storyboards. The sequencing of steps represents the storyboard proper. The notes allow observations to be made about each step that have implications for the "shape and form" of the associated output. The last column lists features that are required of the output if the associated step is to be completed satisfactorily. Collectively, all of the entries that appear anywhere in the last columns of all the storyboards for a particular output represent the fitness-for purpose features of that output.

5.3 The Business Case

The business case is the most significant output from the initiation phase. A business case, if developed through a well-considered process and including reliable investment information, plays a critical part in laying the foundation for a successful project. This section introduces a representative structure for a business case, discusses an example and describes the process by which a funding decision is taken.

Table 5.7 Example of a tentative story board for Shanghai tourism initiative

Project	Project customer:	Output:	Target Outcome:
Shanghai tourist promotion	Tourists from Shanghai	Plaques at points of interest	Increased visitation to the township
Storyboard			
#		Notes	Implied fitness-for-purpose features of plaques
1	Tourist sets out along heritage trail to next plaque		Clearly signposted from heritage trail Easily found from heritage trail
2	Stops at plaque to read		Dual English/Chinese Readable from outside
3	Views features at point of interest	If interest is triggered, may want brochure to take away May want place to sit	Accessible viewing points Brochure display stand Protected seating at some locations
4	On return to Shanghai, describes the experience to acquaintances, friends and relatives	Others are encouraged to make the same trip	Material displayed on plaques should be interesting, attractively displayed and noteworthy

Box 5.6 Outputs from Initiation

A Suggested Template: The Business Case

1. Introduction

   1.1. *Purpose of document.* In most cases this will simply confirm that the purpose of the document is threefold: to support a funding decision, to establish the project's key parameters and to brief key stakeholders.

   1.2. *Overview of the project.* Outline the project by briefly discussing the "shape" of the proposed initiative.

   1.3. *Project appraisal.* Summary of benefits, disbenefits, costs and risk exposure.

2. Business context

   2.1. *Background.* What led to this project?

   2.2. *Rationale and Strategic fit.* Why this particular project at this particular time? How does this project fit into the organisation's strategic or policy framework, and how does it rank amongst competing initiatives?

   2.3. *Organisational impact statement.* What impact will the work of this project have on the organisation (especially demands for internal resources), and how will that impact be managed?

   2.4. *Scenario analysis.* Where appropriate, discuss the "Now", "No" and "Yes" scenarios.

   2.5. *Analysis of options.* What options (if any) were considered for the exercise? How were they ranked? Why is this the preferred option?

   2.6. *Related projects and programmes.* To what projects and programmes is this one related, and how will those links be managed? Three relationships are of interest: projects on which this one depends, projects that are interdependent with this one, projects that will depend on this one.

   2.7. *Assumptions and constraints.* What values of key variables and conditions have been assumed? What constraints have been imposed on the project? What are the implications for the business case of changes in each of these?

3. Project Definition

   3.1. *Statement of scope*

   Project objective. A short statement that answers the question "Why is this project being funded"

List of target outcomes. A simple list, supported below with definitions of each item.
List of committed outputs. A simple list, each item is defined below with a list of critical fitness-for-purpose features.

3.2. *Outcomes definition.* A table in which:
Columns are associated with target outcomes.
Rows with the (seven) attributes used to define a target outcome.

3.3. *Undesirable Outcomes* (optional). List, identify impactees and outline management programme.

3.4. *Outputs definition.* For each output, there is a list of its critical fitness-for-purpose features.

3.5. *Excluded outputs* (optional). Sometimes the scoping process generates expectations of outputs that are eventually culled from the scoping statement. In those cases, it may be helpful to display another list of outputs that are not in scope.

3.6. *Utilisation map*

3.7. *Utilisation storyboards*

3.8. *Financial appraisal.* Analysis of cashflows from the perspective of the funder.

4. *Stakeholder analysis.* Supported with a Stakeholder Register

5. *Project Governance.* Supported with: a governance model and role definitions for key players

6. High-level plan

6.1. *Preliminary workplan.* This includes a schedule of major milestones. Estimates of duration should be based on assumed levels of resourcing and qualified with clear statements about their reliability, typically by showing ranges.

6.2. *Resource plan and budget.* The amount of resources and the budget required to complete the work of the project. These estimates should also be qualified with clear statements about their reliability, typically by showing ranges.

7. Issues and risks

7.1. *Critical risks.* Based on a Risk report.

7.2. *Key issues.* Based on an Issues report.

8. *Recommended approach to planning*
A high level outline of how the planning phase will be approached and managed.

5.3.1 *The Structure of a Business Case*

We propose the following structure for a business case:

A key part of the business case is Sect. 1.3 in Chap. 1 "Project appraisal", which summarises the project's anticipated worth. This analysis is based on estimates that appear later in the document. Estimates are predictions about the future values of variables. Predictions are subject to uncertainty, which will eventually be reflected in errors. Decisions involving an estimated variable require two pieces of information about that variable: the estimate itself and (equally important) a clear indication of the reliability of the prediction. So all quantities derived during initiation *must* be qualified with a statement about the confidence that the champion places on those figures. ("Confidence" in this case relates to the risk of the parameter deviating unfavourably from its desired threshold). Point estimates are to be treated with suspicion because they suggest absolute certainty, a situation that is rare at this point in the life of any project.

A declaration concerning the uncertainty surrounding the estimates in the business case is critically important to the reliability of the decision to proceed. But now, a second trade-off has to be made during initiation, between measuring the accuracy of the estimates that have been provided and the time and cost of that work. There are various ways of qualifying project estimates, three of which are worthy of note here.

- The first involves use of a wordscale for the accuracy of the estimates (such as: poor, low, acceptable, and high). Because of the possible differences in interpretation, this approach does little more than alert decision-makers to quality problems in estimates.
- The second involves an upper and lower bound on each estimate (often based on pessimistic and optimistic scenarios). In practice, it is difficult to know what qualifies as representative pessimistic and optimistic scenarios, and so the ranges implied by the bounds may, in fact be much wider.
- The third uses three values for each estimate: an upper bound, a lower bound and an expected value. This is in fact the basis of some well-worked techniques such as PERT which allow statistical and mathematical analysis to be applied.

While approaches to estimation are discussed more fully in Chap. 6, a comment about the *high-level plan* (that appears as item #6 in the above template) is necessary. Estimates of duration and cost can be developed in either of two ways: by considering experience with similar outputs in past projects (output-based estimation) or by analysis of the work required to create the outputs (work-based estimation). The second approach cannot be used during initiation because a detailed model of the work required on the project is normally not available until the next (planning) phase. During initiation, output-based estimates are normally employed. This approach is quick, but yields figures that are surrounded by wide ranges of uncertainty.

5.3.2 A Business Case Example

The following example of a business case applies the template suggested above to the ICO case study, Project BuyRite. In some places, we have confined the discussion to representative illustrations of the complete material that one would expect to find in a real-life document. Later sections on outputs definition and utilisation storyboards are cases in point.

1. Introduction

   1.1. *Purpose of document.*
   This document seeks funding from the ICO Board to undertake Project BuyRite, an initiative to improve the Company's approach to procurement. The business case also confirms the project's key parameters and serves as a brief for key stakeholders.
   1.2. *Overview of the project.*
   Project BuyRite seeks to re-engineer global procurement practice and support this with appropriate technologies, staff training and organisational change.
   1.3. *Project appraisal.*
   Project BuyRite is a very high yield, high cost, high risk initiative.

2. Business context

   2.1. *Background.*
   Much of ICO's recent growth has been driven by an aggressive programme of international acquisition. The business processes in existence across the Company are, in general, modified versions of those used by various original local operators before they were absorbed into ICO. Procurement, for example, is approached differently in each of the company's numerous centres of operation.
   The recent international benchmarking study has revealed serious problems in this area that are having a significant and unacceptable impact on the Company's financial performance. It is proposed that a standard, best-practice suite of procurement processes should be implemented across ICO.
   2.2. *Rationale and Strategic fit.*
   Project BuyRite gives effect to ICO's recently-approved global development strategy, as outlined in the document "ICO as leader of the international concrete industry". That paper identified procurement, manufacturing and order fulfilment as our three top-priority business initiatives for the next 2 years.
   2.3. *Organisational impact.*
   Project BuyRite will have a significant impact on day-to-day operations because of the need to take some of our best staff out of their business units to work on the project for periods of between one and 2 years.
   The resulting operational staffing shortfall will be addressed through the appointment of contractors. Our analysis indicates that this will cause

procurement costs to rise somewhat over the course of the project and that there may also be some slow-down in sales growth as the project nears completion (due to an inevitable decline in procurement performance).

2.4. *Related projects and programmes.*

Projects dependent on BuyRite. A business case is to be prepared next year for Project BatchRite (which will re-engineer our manufacturing processes). Project BatchRite is heavily dependent on Project BuyRite in that the redesign of our inbound logistics will be constrained by what is done with procurement.

It is proposed to establish an Inbound Logistics Working Party (IL/WP) as a reference group in the Project BuyRite governance model. All procurement decisions that impact manufacturing will be referred to this reference group for review. If, as planned, Project BatchRite is approved before BuyRite finishes, then the ILWP will also serve as a reference group for Project BatchRite.

Projects interdependent with BuyRite. The Global Finance Office has just started work on "FINAC" (a project to improve accounting practice across the Company) and, as part of that exercise, replace the existing finance and accounting software system. It is proposed to establish an Accounting Operations Working Party (AO/WP) as a reference group in the Project BuyRite governance model. All procurement decisions that impact accounting will be referred to this group for review.

Projects on which BuyRite depends. A decision was taken 12 months ago to outsource the bulk of IT services to Technical Infrastructure Management Services Inc (TIMS). The plan for this exercise has the transfer of IT operations in Australia taking place about the same time as BuyRite will be implemented. It is proposed that an adviser be appointed from BuyRite to the outsourcing project to ensure that the transfer of Procurement's infrastructure to TIMS is timed appropriately.

2.5. *Assumptions and constraints.*

Assumptions. It is assumed that no major acquisitions will be made during Project BuyRite. If that happened, it may prove necessary to slow the project so that senior Procurement staff could be made available to support integration of the new business unit into ICO.

Constraints. A subsidiary of ICO will move out of the global HQ building in 4 months. That will provide adequate long-term accommodation for the BuyRite team. Until then they will have to arrange temporary offices elsewhere.

3. Project Definition

3.1. *Statement of scope*

Project objective. To achieve world class performance in ICO's procurement operations.

List of target outcomes.

- Reduced procurement costs
- Reduced payment times to suppliers

List of committed outputs.

- New procurement processes
- New procurement policy and procedures manual
- A restructured procurement unit
- Enabling applications systems and new technical infrastructure
- A panel of preferred suppliers
- Programs of accredited professional development for staff

3.2. *Outcomes definition.*

Attribute	Target outcome title	
	Reduced procurement costs	Reduced payment times to suppliers
Description	Expressed as the average cost per purchase order issued over a quarter (including consumables and labour)	Expressed as the 95 percentile time to settle a clean supplier invoice. Calculated over a quarter
Measure	Dollars (per purchase order)	Days (to settle)
Target	A reduction of 25%	A reduction of 50 days
Source/method	As shown in proposed annual procurement performance report, based on data from the new management accounting system	As shown in proposed annual procurement performance report, based on data from the new management accounting system
Achievement date	Target to be realised by Q4 2011	Target to be realised by Q1 2012
Person accountable for realising the target outcome	Nancy Palmer, National Procurement Manager	Nancy Palmer, National Procurement Manager

3.3. *Outputs definition.*

Each output identified in the scoping statement is defined here by listing its fitness-for-purpose features. [*In this illustrative business case, only one output is defined in this way*].

New procurement policy and procedures manual.

- Available on-line.
- Structured as hypertext (for ease of navigation).
- To include comprehensive details of all new procurement processes, as interactive process models.
- Subject to a quarterly review/revision process.
- To be covered in training and development programme for Procurement staff.
- To include guidelines for vendor management.

3.4. *Utilisation map.*

Output	Target outcome	
	Reduced procurement costs	Reduced payment times to suppliers
New procurement processes	Procurement staff Suppliers	Procurement staff
New procurement policy and procedures manual	Procurement staff	Procurement staff
A restructured procurement unit.	Procurement staff	Procurement staff
Enabling applications systems and new technical infrastructure	Procurement staff Suppliers	Procurement staff
A panel of preferred suppliers	Procurement staff	
Programs of accredited professional development for staff	Procurement staff	Procurement staff

3.5. *Utilisation storyboards*

Each project customer identified in the utilisation map links one output to one outcome. For each link of this kind (of which there are 13 above) there will be an associated storyboard describing how utilisation of the output generates the target outcome. [In this illustrative business case, only one storyboard is shown].

Project customer: Procurement staff	Output: New procurement processes	Target Outcome: Reduced payment times to suppliers
Storyboard # Project customer utilisation step	Notes	Implied fitness-for-purpose features of new process
1 Procurement staff confirm that 3-way matching is OK.	3-way matching reconciles: supplier invoice, purchase order and goods receipt	3-way matching to be done automatically
	Procurement staff only become involved if there is a problem	Special orders (such as special engineering equipment) require a delivery certification process involving originator of order A successful match triggers automatic scheduling of payment to supplier
2 Procurement staff are notified automatically of payments that exceed preset threshold	Procurement staff attend to late payments	Process allows for: Warning to staff of delayed payments Manual follow-up of delayed payments

3.6. *Excluded outputs*

Early work on the business case identified two candidate outputs: a performance bonus scheme for procurement staff and a new office for the new procurement unit. Both were subsequently excluded from the scope of BuyRite.

3.7. *Undesirable outcomes.*

It is anticipated that Procurement staff will find aspects of project BuyRite disruptive, and the resulting changes to the procurement environment challenging. This could lower staff morale and even lead to resignations. Two steps are being taken to manage this:

- The project governance model for the project recognises a role for a Procurement staff reference group.
- The training programme in the new processes will acknowledge this situation.

3.8. *Financial appraisal*

A complete analysis of the cashflows associated with Project BuyRite appears as an Appendix to this document. This indicates an NPV of $2.5M.

4. Stakeholder analysis

A full stakeholder register would normally be shown here, but for this example we summarise the details that apply to just one stakeholder, the Procurement staff. [These details are displayed vertically here, whereas a typical stakeholder register would show them as a row].

Attribute	Entry
Name of candidate stakeholding entity	Current procurement staff
Nature of stakeholding	They are the subject of a significant change programme Their input to the re-engineering of ICO's procurement processes is critical to project success
Classes of spontaneous stakeholding	Impactees
Issues arising from stakeholding	There will be varying degrees of support for, and resistance to, change
Commissioned stakeholding	All procurement staff will be invited to join a Reference Group
Engagement strategy	Include in Project Governance (as members of a Reference Group) Include in the communications plan for project BuyRite (conduct quarterly review workshops, send copies project newsletter and keep updated on project website)

5. Project Governance Model

[This section would normally include not only the project governance model as a diagram (like that shown in Fig. 5.7), but also the Terms of Reference for all entities identified in that model. Here we show only the Charter for the Steering Committee].

5.1. *The Steering Committee*
      Name of PGM role. Project BuyRite steering committee

> *Objective of role.* The steering committee is to guide the project towards successful realisation of the business case.
>
> *Outputs from role.*
>
> - Guidance to Project Manager
> - Decisions about course of the project
> - Resolved issues
> - Mitigated risks
> - Acknowledgement of deliveries by project team
>
> *Core activities and frequencies.* Monthly meetings at which: the project manager presents on the status of the project, decisions are taken, action items accepted and guidance/instructions issued to the project manager. The steering committee is also to undertake an outcomes close-out workshop as soon as target outcomes have been secured.
>
> *Membership/leader.*
>
> - Nancy Palmer (Chair)
> - Charles Edwards (CEO)
> - Owen Oliver (COO)
> - Catherine Farnham (CFO)

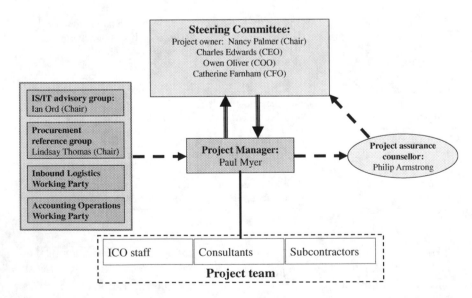

Fig. 5.7 The project governance model of project BuyRite

Where in PGM does this role report? The steering committee reports (through Nancy Palmer as project owner) to Charles Edwards as funder.

Review of role. The role of the steering committee will be reviewed every quarter by Philip Armstrong (Project assurance counsellor). This is to take place in a special meeting/workshop.

Term. The steering committee will come into existence immediately and continue its role until target outcomes have been secured.

6. High-level plan

The following schedules of milestones and costs were obtained from output-based estimates provided by our process re-engineering consultants—based on their experiences with similar projects both here and overseas during the past 10 years:

6.1. *Preliminary workplan.*

#	Activity	Start date	Finish date
1	New procurement processes:		
	Development	1 February	30 April
	Approval	1 May	15 May
2	New procurement policy and procedures manual:		
	Development	1 February	30 April
	Approval	1 May	15 May
3	A restructured procurement unit:		
	Development	1 February	31 May
	Approval	15 June	30 June
4	Enabling applications systems	1 February	31 July
5	New technical infrastructure	1 February	31 July
6	A panel of preferred suppliers		
	Development	1 June	15 June
	Approval	16 June	30 June
7	Programs of accredited professional development for staff	1 April	31 July

6.2. *Resource plan and budget.*

#	Output	Cost ($)
1	New procurement processes	10,000
2	New procurement policy and procedures manual	50,000
3	A restructured procurement unit	20,000
4	Enabling applications systems and new technical infrastructure	150,000
6	A panel of preferred suppliers	10,000
7	Programs of accredited professional development for staff	50,000
	Total	290,000

7. Issues and risks

   7.1. *Critical risks.* Based on a Risk report.
   A full risk register would normally be shown here, but for this example we
   summarise the details that apply to just one threat, related to Lindsay
   Thomas's availability. [The details for this risk are displayed vertically
   here, whereas a typical risk register would show them as a row].

Attribute	Entry
#	R17
Threat	Lindsay Thomas becomes unavailable during project
Pre-likelihood	Distinct chance (from standard wordscale)
Impact	Benefits delayed
	Costs increased
Pre-Severity	Grave (from standard wordscale)
Pre-Risk Exposure	This is a Grade E threat
Risk mitigation plan (RMP)	(P) Accelerate the work on process analysis by
	employing world's top process analyst
P = Preemptives	(P) Arrange a significant completion bonus to LT
C = Contingencies	(C) Appoint another senior member of the
	Procurement Staff as an understudy
Post-likelihood	Remote chance (from standard wordscale)
Post-Severity	Significant (from standard wordscale)
Post-Risk Exposure	This becomes a Grade H threat
Effectiveness of proposed RMP	3/11
Cost of proposed RMP	$150,000

   7.2. *Key issues.* Based on an Issues report.
   A full issue register would normally be shown here, but for this example
   we summarise the details that apply to just one issue, related to office
   accommodation for the project team. [The details for this issue are displayed vertically here, whereas a typical issue register would show them as
   a row].

Attribute	Entry
#	I42
Issue	The recent sale of ICO's concrete additives R&D business will
	free an office that could be used by the BuyRite team
Importance	High (from standard wordscale)
Status	Active
Notes	The current lease was negotiated at very attractive rates and
	runs for another 2 years. Some fit-out would be required
Assigned to	Pasquale Mataro (ICO's Property Manager)

5.3.3 Judging a Business Case

Project appraisal is undertaken separately (and for different reasons) by the funder, the project-owner-designate and the project-manager-designate. All assess the business case, but from different perspectives. The funder's appraisal is centred on the question "Do I believe that the project defined by the business case represents an appropriate investment?". The project owner wants to know if the business case can be realised (with an acceptable probability). By way of contrast, the project manager's appraisal is concerned with the question "Do I believe that the project is feasible?".

The funding decision bears further discussion. Acceptance of the business case requires that the candidate project be ranked against others in terms of "attractiveness". The attractiveness of a project is determined by both its anticipated worth (as a threshold) and the achievability of that threshold. Anticipated worth involves estimates of three sets of variables (benefits, disbenefits and costs), and so the funder needs to know not only the predicted thresholds of values for those variables, but also the reliability of the predictions. The reliability of a project's anticipated worth is an expression of its risk. Low reliability of estimated variables implies a high level of risk that threshold value of the project will not be achieved, while a high degree of reliability in those figures implies a low level of risk.

In summary, to determine the anticipated worth of a project we need to know about the flows (over time) of:

1. *Benefits*. A benefit is a "flow of value" (as judged by the funder) to a project stakeholder arising from achievement of target outcomes.
2. *Disbenefits*. A disbenefit is a "flow of value" (as judged by the funder) away from a project stakeholder in the course of the project.
3. *Costs*. The funds required to produce and maintain the project's outputs.

In addition, we also need to know about the risk that thresholds for all three values will not be achieved. For example, in the case of Project BuyRite, before approving the project, Charles Edwards (as prospective funder) will want to know (amongst other things):

1. *Benefits*. "What flows of reductions in procurement costs that I can expect?"
2. *Disbenefits*. "What will be the extent of any fall in morale amongst procurement staff?"
3. *Cost*. "How much will I need to invest in the project?", "Will there be any ongoing extra operational costs?"
4. *Risk*. "How reliable is all this information?"

Investment theory offers a framework to guide the appraisal process, however many of the techniques of investment theory cannot be used directly in project analysis for two reasons: worth usually involves non-financial units of measurement and only "downside" risk is considered.

Box 5.7 Illustration of Concept: Project Appraisal based on "Attractiveness"

Project BuyRite Case Study

While the decision to approve the project will be based on all the information contained in the business case, the analysis of benefits, disbenefits, costs and risks is of particular interest to the CEO (Charles Edwards):

1. *Benefits.* Defined target outcomes are recognised as the drivers of project benefits. The objective of improving the performance of the Procurement Department has been translated into two target outcomes—reduced procurement costs and increased speed of invoice settlement. The first of these represents a benefit to ICO as it stands. As stated, the second represents a direct benefit to suppliers. Although there may also be an indirect benefit to ICO from reduced payment times to suppliers (perhaps in the form of improved early payment discounts), this effect is unstated and ignored in this case.

   Because it is expected that, once secured, the target outcomes will continue to be generated at a constant monthly rate, they are understood to take the form of flows in time.

2. *Disbenefits.* In Project BuyRite, a plausible disbenefit relates to the downsizing of the Procurement Department. Staff will quickly become aware of this possibility and, understandably, are likely to experience a period of low morale. It could even cause key staff to leave. Although these factors are certainly not large enough to make the project unattractive, they are, on their own, undesirable effects, and so "reduced morale" becomes a project disbenefit.

   Analysis by Paul Myer and Nancy Palmer suggest that, even with a sophisticated stakeholder engagement plan, there will be an appreciable fall in procurement staff morale for the first 9 months of the project.

3. *Costs.* Using a detailed WBS and workplan (not shown here), the costs for Project BuyRite have been estimated as $750,000, taking the form of equal quarterly outlays of the period covered by the project workplan.
4. *Risks.* The risk register as included in the business case indicates that the exercise is, overall, of medium risk (after accounting for proposed mitigation programmes).

   The conclusion drawn from all this is that the project is very attractive and should proceed.

A project will be approved for execution only if the funder judges the notional value of target outcomes on one hand to be significantly greater than the combined notional value of undesirable outcomes and outlays on the other. An illustration of project appraisal for the Project BuyRite follows.

5.3.4 Accepting the Business Case

Because the business case is such an important document, the funder would normally seek the views of other key stakeholders before formal acceptance. Throughout the initiation process, the project champion should brief the funder about the project as working drafts of the business case become available. This phase ends with a presentation by the champion of the business case to the funder.

Significant projects may well involve special presentations of the business case by the project champion to other senior managers and selected key stakeholders.

Tabling of a business case can result in three possible decisions:

1. *Accept.* The project is approved and progresses to the planning phase. The business case as it stands becomes the foundation for the eventual project plan, although the funder may first want to highlight certain elements in the project approval document, such as:

   - Agreed project objective, outcomes and outputs
   - Approved budget and resources
   - Expected timelines
   - Expected level of quality
   - The name of the project owner
   - The name of the project manager

2. *Reject.* The proposal is not accepted and the idea is abandoned (or shelved).
3. *Rework the business case.* Often the funder will require more information before the project is approved. In this case the business case will have to be reworked. If the required changes are minor, an in-principle approval may be granted, whereby planning may start as soon as the revised business case is tabled.

It should be noted that acceptance of a business case is always an in-principle funding decision because one of the outputs from the planning phase is a modified business case which allows the original funding decision to be revisited and confirmed.

5.4 Appraising Project Risk

The interest that various key players have in project risk arises from the following observations:

- Most projects are exposed to chance events that have damaging consequences.
- The damaging effects of these events can, in extreme cases, lead to project failure.
- Many events of this kind are identifiable in advance.
- The damaging events to which a project is exposed vary in terms of their "importance", and can be ranked accordingly. Those with high importance appear to demand action, while those with low importance can, in many instances, be ignored.
- We can take action against certain risks that can reduce their importance and hence the amount of attention that they subsequently demand.
- The costs of taking action against some risks appear low when compared with the extent to which their importance is reduced.

Under these conditions we appear compelled to act against the risk. (One could even argue that a failure to act would constitute a form of professional negligence).

Here we provide a broad, high-level overview of project risk and its management. The approach we propose below differs from conventional wisdom in two respects: it is peculiar to the project environment, it confines attention to downside impacts. A more comprehensive discussion of the underpinning concepts, tools and techniques appears in Chap. 6.

5.4.1 The Level of Project Risk

Risk arises from uncertainty about certain events surrounding the project that have a damaging impact on the project's worth. Risk can be viewed, therefore, as uncertainty about the achievement of a project's anticipated worth. Events that would lower a project's worth are called "threats". By implication, if an event can have no impact on a project's worth then it cannot represent a threat to that project. Threats can be ranked by their risk exposure, a qualitative measure of their "importance" derived from the likelihood of the damaging event occurring and the amount of damage suffered if it does occur.

The "riskiness" of a project is gauged by the risk exposure arising from all the threats to which the project is exposed. Consider two projects "X" and "Y" that are identical in all respects except that "Y" has a higher overall risk exposure to failure than "X". Clearly "X" is a more attractive investment opportunity than "Y". This argument implies that if projects are to be ranked in terms of their attractiveness for funding, then we must consider both their worth and their riskiness. While such a conclusion is completely consistent with the accepted principles of investment theory, most of the analytical tools used in that discipline cannot be applied to project funding decisions for two reasons:

- We are, in general, limited to qualitative measures of worth and riskiness.
- In investment theory, risk is seen as the uncertainty of *exceeding*, as well as *falling short* of some expected value. Here we are concerned only with falling

short of a threshold, a stance that is consistent with satisficing (rather than optimising) behaviour.

We identity this particular approach to risk, in which we are concerned only with the possibility of not reaching the satisficing thresholds that define the project's worth, as asymmetric risk management. Unlike the more common approach (as found, for example, in investment analysis), we place little value on over-performance, but very high value on underperformance. Amongst other things, this enables us to treat risk as intrinsically undesirable and so risk aversion is implied, we do not have to assume it.

Risk management is a formal process by which threats are identified, analysed and, where appropriate, mitigated. Risk mitigation programmes are based on actions that lower the risk exposures of selected threats. Such actions are of two kinds:

- Preemptives reduce the likelihood of the threat emerging in the first place.
- Contingencies reduce the severity of the damaging impact if the threat is realised.

Take the threat "project manager leaves" for example. Preemptives include: put him/her under contract, offer a "golden handshake" (a completion bonus) or increase his/her remuneration. A contingency would be to have a deputy trained and ready to step in if required. A risk mitigation programme results in changes to the risk exposures of certain threats and so the risk management process has to accommodate two values of likelihood, severity and risk exposure; one representing the values of these three variables in the absence of any mitigation and another representing the values assumed after mitigation. These are identified as the *pre-* and *post-* values respectively. The application of pre- and post- values for risk related variables is discussed more fully in Chap. 6.
Because they involve additional resources, mitigation programmes have the effect of lowering a project's risk exposure by increasing its costs. The funder (through the project owner) must, therefore make a trade-off between cost and risk.

5.4.2 The Effect of Risk on Project Appraisal

The attractiveness of a project to a potential funder is determined by a combination of worth and riskiness. The higher the worth of a project the higher the levels of risks that a funder is prepared to accept. Similarly, the approval of a "low worth" project, usually requires very low level of risk to be attractive for a funder.

Ranking project alternatives requires that a judgement be made about the extent to which increasing worth is compensation for increasing uncertainty about whether or not the extra worth will be realised. While it is obvious that projects with a high worth and low risk are more desirable than those with a low worth and

Fig. 5.8 A number of
projects ranked by
attractiveness

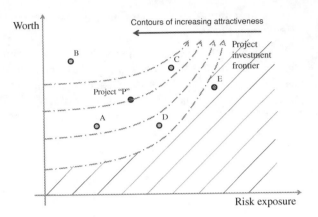

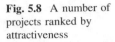

Fig. 5.8 A number of projects ranked by attractiveness

high risk, the situation with other combinations of worth and risk requires deeper
analysis. Figure 5.8 shows conceptually, how projects can be plotted using two of
the fundamental parameters contained in their business cases: worth and risk
exposure. Levels of attractiveness are indicated by contours (of which the project
investment frontier introduced in Chap. 3 is one). Because Fig. 5.8 is concerned
only with downside risk, contours must rise monotonically (that is, they must have
a "positive slope"). Consequently, attractiveness rises towards the upper left. If
"P" is some reference project, and the other points indicate competing alternatives, then the rankings will be decided by the "slopes" of the contours. As
displayed here, in increasing order of attractiveness, the proposed initiatives would
be ranked as: D, A, P, C, B. (E falls "beneath" the project investment frontier and
hence is unsuitable for funding regardless). Looked at another way, B is the best
funding opportunity because it is both higher in worth and lower in risk than all the
others. This means that no matter how the contours are sloped, B will always
"win". In the language of decision theory B is said to "dominate" all the others.
Consider a situation where there is no B. Given the slopes of our contours, while C
would now be the best funding option, it does not dominate P, A or D and so its
ranking is sensitive to the "slope" of the contours.

5.5 Summary

In this chapter, we have described the work required to get a project under way,
with enough supporting analysis and information so that a reliable decision can be
made not only about funding, but also about accepting the ongoing roles of project
owner and project manager. When the business case is accepted, it provides the
foundation for a very detailed model of the work required to produce the project's
outputs. That model is called the project plan, the subject of the next Chapter.

Chapter 6
Planning a Project: The Roles of the Key Players

6.1 An Outline of Project Planning

The planning phase of a project has three objectives: to assemble a detailed model of the work required to produce the deliverables identified in the statement of scope, to reconfirm the earlier in-principle funding decision and to establish the environment in which the project will be executed. The foundation for planning is the business case.

6.1.1 The Need for Planning

Because project execution is a process (albeit a very large one), it requires a "script" that describes, in appropriate detail, how the work involved is to be carried out. While the business case includes a high-level outline of *what is to be done*, it does not describe *how that work is to be performed*. A planning phase is introduced into the overall structure of a project to prepare such a script. Project planning involves considerable analytical effort. It also demands significant resources and appreciable elapsed time. If the quality of project plans was unrelated to eventual levels of success then, clearly, planning would be a waste of time and resources. However, a very strong case can be mounted in support of the claim that the quality of a project plan is a significant determinant of eventual success (Johnson, Karen, Boucher, & Robinson, 2001; Pinto & Slevin, 1989; Zwikael & Globerson, 2004; Zwikael & Sadeh, 2007). As a result, "Failing to plan is planning to fail" has become something of a mantra among project management practitioners (McNeil & Hartley, 1986). Not surprisingly, planning figures prominently amongst the key factors of project success (Dvir & Lechler, 2004; Johnson et al., 2001; Pinto & Slevin, 1989). As Dvir, Lipovetsky, Shenhar & Tishler (2003, p. 89) put it: "In fact, although planning does not guarantee project success, lack of

O. Zwikael and J. Smyrk, *Project Management for the Creation of Organisational Value*, DOI: 10.1007/978-1-84996-516-3_6, © Springer-Verlag London Limited 2011

planning will probably guarantee failure". Consequently, many believe that a threshold of planning is required, in which the level of effort and amount of detail depends on the project in hand.

Although most scholars and practitioners agree that planning improves project success, some have added words of caution, especially when it comes to

Box 6.1 Practical Considerations

The "We don't have Time to Plan!" Paradox

Most project managers will face, at one time or another, an argument supposedly "proving" that planning actually raises the risk of project failure! A senior project stakeholder who genuinely (but mistakenly) believes that planning wastes precious time will present an argument along the following lines:

1. *"The project must be complete by (say) the end of November"*. This is usually supported with references to declarations made by various key players (such as the funder) or authorities (such as a regulatory agency). "Complete" usually means the date by which all the outputs must be delivered.
2. *"Since we are now at (say) the beginning of July, this gives us 20 weeks to do the work"*. Mathematically this is indisputable, there are indeed 20 weeks between the beginning of July and the end of November.
3. *"If you do any planning that will, effectively, divide the 20 week timeframe into two parts: a planning part and an execution part"*. That is certainly true—we are proposing a period of planning followed by a period of "productive" work.
4. *"The longer the planning phase, the shorter the execution phase (because together, they must be completed in only 20 weeks)"*. Again the logic of that statement is clear.
5. *"The shorter the execution phase, the greater the risk of not meeting the end-of-November deadline"*. It is hard to argue that the shorter the timeframe the lower the risk, therefore, this statement seems to be true as well.
6. *"Therefore, to minimise the risk of failure, we must allow the maximum time for execution, therefore, don't plan, just do it!"*. But we know that without planning the project will almost certainly fail. Something is clearly wrong, but what?

This sort of (totally flawed) argument can be very difficult to detect and even more difficult to counter, especially for inexperienced project managers. The discussion about infeasible projects (in Box 6.4) holds the key to managing situations like this.

"exploratory projects". Andersen (1996) claims that too much planning can curtail creativity. In regard to research and development (R&D) projects, Dvir et al. (2003) have stated, "there is no correlation between the implementation of planning procedures in the project and the various success dimensions". They also suggest that the selection and application of particular planning toolsets is of little importance. Yet, in practice, project managers significantly improve their project plans in the presence of high risk (Zwikael & Sadeh, 2007). Such research results are intriguing because they seem to weaken the case for planning in R&D projects.

There are a number of explanations for this apparent inconsistency of views about the effectiveness of project planning. One is that the horizon over which R&D project plans remain valid is much shorter than for other types of exercise. Such projects may yield to an alternative "progressive" approach in which the planning horizon is broken into discrete stages and plans are confined only to the immediate stage. This approach is similar to, what is sometimes called, "rolling wave planning". Another explanation is that in high risk projects, (when uncertainty is significant and the quality of estimates is correspondingly poor), planning is difficult and hence project plans are of somewhat limited value. In summary, Dvir and Lechler (2004) suggest "contextual settings" for planning in different projects, while Zwikael and Globerson (2006) suggest an exclusive focus on planning processes in various project scenarios. We have also offered some observations about such projects by suggesting that they be structured as a "staged" programme (in Sect. 4.6).

6.1.2 The Structure of the Planning Phase

The planning phase of a project is broken into two parts: planning proper and set-up. The bulk of this chapter is concerned with the first of these, but on larger projects the second can represent a significant exercise in its own right. It is during set up that the project environment is created. This work is highly specific to the particular project. On small exercises it may involve little more than arranging the first meetings of the various forums identified in the project governance model, while at the other extreme, set up may require large-scale activity, such as the establishment of temporary offices, visas for overseas team members, leasing a fleet of cars and acquisition of office equipment.

Before setup can begin, the project plan (together with the modified business case) must be tabled and approved, as suggested in Fig. 6.1.

Two activities are common to all project environments, regardless of the size of the initiative being undertaken:

- Establish the project governance model, which involves tasks such as: prepare terms of reference/charters/job descriptions, appoint people to defined roles, draft supporting contracts/memoranda of understanding, induct appointees into the project.

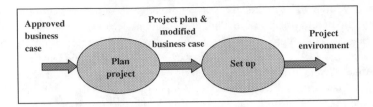

Fig. 6.1 The project planning processes

- Create project infrastructure, especially accommodation and computing/communications facilities.

6.1.3 The Outputs from Planning

Planning gives rise to two significant sorts of output: new baseline documentation and (later, when set up is complete) the project environment itself. The project planning document (usually identified simply as "the project plan") is one of two baseline documents on which the eventual conduct of the project is based, the other being the business case. The term "baseline" indicates that everything that is done on the project must be carried out in accordance with these documents.

A project plan is obtained, in effect, by augmenting the business case with additional detail of four different kinds:

- *New information* that was not included in the business case (and which in many cases, will not have been available at that time). Examples come readily to mind, such as: quality management and communications plans.
- *Revised values* (where necessary) for certain parameters that had been set in the business case (for example, a more reliable estimate of total financial outlays).
- *Expanded detail* about certain aspects of the project that may have been only outlined in the business case (project governance is a case in point).
- *Additional entries* in the registers that first appeared in the business case (for example, new risks that have been identified since the original business case was tabled).

While planning is primarily concerned with the assembly of a project plan, it inevitably involves updating or revising certain parts of the business case, and so this phase of a project has two major outputs: a (new) project plan and a (modified) business case, as suggested in Boxes 6.2 and 6.3 below.

Some observations are appropriate on two sections of these templates that are not covered elsewhere in this chapter:

- *The business impact statement* makes explicit the internal resourcing issues raised by the project. In particular it addresses the question "How will the demands on internal staff be met?". Too often the glib response "Work smarter not harder" is code for "Work harder and longer". When such an approach is

Box 6.2 Outputs from Planning

A Suggested Template: The Project Plan

1. Introduction

   1.1. *Purpose of document.* Confirms that document seeks to: augment the modified business case, provide the information on which a reliable decision to start work can be based, brief key stakeholders and outline what happens next.
   1.2. Relationship to the modified business case.
   1.3. A summary of the structure of the document.
   1.4. Key points raised in major sections.

2. Stakeholder management.

   2.1. Stakeholder Register.
   2.2. *For each stakeholder:* A detailed engagement plan.
   2.3. Communications plan.

3. *Project governance.* Discussion of recommended project governance model, membership roles, responsibilities and management arrangements.

4. Workplan

   4.1. Work breakdown structure.
   4.2. *Outputs.* A simple list of all outputs. For each output:

   - Critical fitness-for-purpose features
   - Quality criteria for each fitness-for-purpose feature

   4.3. Gantt chart.
   4.4. Schedule of milestones.
   4.5. *Quality management.* An approach to quality assurance and control for the whole.
   4.6. *Project management and reporting.* Proposed schedules of meetings for key forums. Description of project monitoring arrangements and the structure/format of periodic reports.

5. Resource plan.

   5.1. Schedule of (financial) outlays.
   5.2. Human resource plan.
   5.3. *Accommodating the project within the organisation.* Discussion of how the work of the project will impact the organisation and how those impacts will be handled. Complete details of how deployment of operational staff will be handled.

6. Risk and issue management:

  6.1. Risk register.
  6.2. Issue register.

Box 6.3 Outputs from Planning

A Suggested Template: The Modified Business Case

1. Introduction

  1.1. *Purpose of document*. Confirms that document seeks to: revise the previous version of the business case, update the project's key parameters, reconfirm the original funding decision, and brief key stakeholders.
  1.2. *Modifications to the original business case*. Summarises the major differences between the original business base and this modified version.
  1.3. *Overview of the project*. Outline the project by briefly discussing the "shape" of the proposed initiative.
  1.4. *Project appraisal*. Summary of benefits, disbenefits, costs and risk exposure.

2. Business context

  2.1. *Background*. What led to this project?
  2.2. *Rationale and Strategic fit*. Why this particular project at this particular time? How does this project fit into the organisation's strategic or policy framework, and how does it rank amongst competing initiatives?
  2.3. *Scenario analysis*. Where appropriate, discuss the "Now", "No" and "Yes" scenarios.
  2.4. *Analysis of options*. What options (if any) were considered for the exercise? How were they ranked? Why is this the preferred option?
  2.5. *Related projects and programmes*. To what projects and programmes is this one related, and how will those links be managed? Three relationships are of interest: projects on which this one depends, projects that are interdependent with this one, projects that will depend on this one.
  2.6. *Assumptions and constraints*. What values of key variables and conditions have been assumed? What constraints have been imposed on the project? What are the implications for the business case of changes in each of these?

3. Project Definition

   3.1. Statement of scope
      Project objective. A short statement that answers the question "Why is this project being funded"
      List of target outcomes. A simple list, supported below with definitions of each item.
      List of committed outputs. A simple list, each item is defined below with a list of critical fitness-for-purpose features.
   3.2. *Outcomes definition.* A table in which:
      Columns are associated with target outcomes.
      Rows with the (seven) attributes used to define a target outcome.
   3.3. *Undesirable Outcomes (optional).* List, identify impactees and outline management programme.
   3.4. *Outputs definition.* For each output, there is a list of its critical fitness-for-purpose features.
   3.5. *Excluded outputs (optional).* Sometimes the scoping process generates expectations of outputs that are eventually culled from the scoping statement. In those cases, it may be helpful to display another list of outputs that are not in scope.
   3.6. Utilisation map
   3.7. Utilisation storyboards
   3.8. *Financial appraisal.* Analysis of cashflows from the perspective of the funder.

4. *Business case material that now appears in the project plan.* What follows is a simple list of the sections that formed part of the original business case, but have since been significantly updated and now form part of the project plan. This includes: .

- Stakeholder analysis
- Project Governance Model
- Organisational impact statement
- High-level plan
- Critical risks
- Key issues

floated it should automatically trigger the entry into the risk register of a threat "Staff unable to meet the demands on their time from both the project and business operations". The only options for meeting incremental demands on staff from new projects are: apply any (fortuitous) existing surplus capacity, delay other work until the project is over, backfill staff with temporary appointments to business operations or arrange for staff to work longer hours. It is important that the preferred strategy be made explicit, otherwise it will default to the last option, possibly putting the project at risk of failure.

- *The quality management (plan)* outlines the way that outputs will be delivered fit-for-purpose. This covers: the quality issues faced with each output, the standards that will be employed for each output and the proposed mechanisms for quality assurance and quality control. Quality assurance is generally accepted to mean procedures that are incorporated into the production process to lower the likelihood of an output emerging unfit-for-purpose. Quality control, on the other hand, refers to testing-type procedures performed on outputs as they emerge from production, but before they are delivered (so that faulty products can be detected and returned for rectification).

It is clear from these two templates that significant changes may be forced on the original business case by planning, both in structure and content. Under normal conditions, however, planning would not result in any significant changes to the underlying thrust of the business case.

6.1.4 An Iterative Approach to Planning

Just as assembly of the initial business case demands an iterative approach, so too does the project plan. As additional information is discovered about one element of the plan, it will have implications for what was uncovered in earlier work on other elements. The spiral that was introduced into Chap. 5 (and used to guide initiation) is readily adapted to serve a similar purpose in planning.

Figure 6.2 shows how the general structure of the spiral approach to project planning relates to (and differs from) that adopted for initiation. A number of points should be noted:

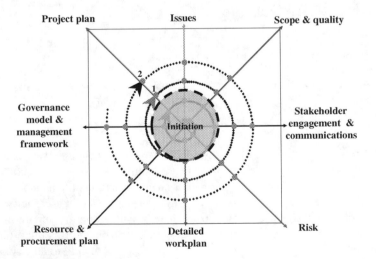

Fig. 6.2 The spiral approach to project planning

- The arms relate to those that appeared in the earlier version and are recognised as project elements (see Sect. 4.2), but now have titles that reflect a planning orientation. "Scope" for example, becomes "scope and quality", consistent with the need for a comprehensive list of fitness-for-purpose features for each output. As was the case with initiation, seven of the arms are associated with project elements, while the eighth is concerned with packaging these elements (in this case) into a project plan.
- The diagram has an "inner" region (represented by a "greyed-out" circle because this work has already been covered during initiation) and an "outer" region (of relevance to the planning phase). The spiral begins just outside the inner (initiation) circle and again progresses outwards clockwise. Each iteration (cycle) adds information along each arm.
- Each point of intersection between the spiral and an arm represents a working session on the project plan.

6.2 The Project Manager

In many cases, the project manager personally undertakes much of the work involved in planning. The reason for this is that not all other project stakeholders (especially project team members) will have been assigned to their project roles at this point. This fact alone, if nothing else, should encourage project managers to prioritise their work. Following a discussion about the responsibilities of the project manager during planning, this section identifies certain important activities ("Critical success planning processes") which should receive special attention. Because they are also of particular concern to key players outside the project team, three tools that require a special focus during planning are described later in this section: the work breakdown structure (WBS), the Gantt chart and the project's cost estimates.

6.2.1 The Responsibilities of the Project Manager

The project manager leads the planning phase (in close consultation with the project owner) and, depending on the size of the exercise, may be supported by others (some of whom could well form the nucleus of the eventual project team). During planning, the project manager will normally be responsible for:

- Deciding on the structure of the plan, guided where appropriate, by the project management standards, conventions and methodologies adopted by the performing organisation.
- Selecting an appropriate toolset.
- Assembling a team to support planning activity.

- Negotiating a timeframe and (if necessary) a budget for the planning phase.
- Managing the planning processes.
- Quality-assuring the emerging project plan.
- Delivering the project plan to the project owner.

As indicated by the study presented in Appendix C, most project managers invest considerable effort in planning. In what follows we discuss the involvement of the project manager in the assembly/production of core elements of the project plan and the modified business case.

6.2.2 Critical Success Processes During Project Planning

"Critical Planning Processes" (CSP), introduced in Sect. 3.5, are those processes that contribute significantly to project success. Here we discuss the most important of these for the planning phase of projects.

The research described in Appendix C, ranked 16 planning processes (identified in the PMI's PMBOK), according to their contribution to project success. Table 6.1 shows the results of this work, together with the PMBOK's knowledge area to which each process belongs.

The four most important of these (according to their impact on project success) merit some comment.

1. *Staff acquisition.* It is generally accepted that the quality of human resources is an important factor in the quality of the work they perform. Specifically, in the project context, it is important that staff members have the relevant knowledge of, experience with, and interest in the specific project to which they are assigned.
2. *Project plan development.* This process involves the assembly of an overall plan for the project, which is based on the integration of several planning processes related to duration, time, cost, risk and others.
3. *Cost estimation.* Reliable estimates of outlays and internal human resources allow for achievable budgets which then have a lower likelihood of being exceeded. Although accuracy in estimates for critical project variables is desired and sought during planning, various factors will limit the reliability these parameters. It is essential that each estimate produced at this time is qualified with a clear statement about its reliability, especially if there is a high level of uncertainty surrounding the figure provide.
4. *Activity definition.* Proper identification of a project's activities is one of the most critical planning processes to be performed by the project manager. This finding makes intuitive sense, since if an activity is left out during the planning phase, its late inclusion afterwards may have a strong negative impact on various aspects of the plan, especially the schedule and budget.

Box 6.4 Illustration of Concept: A Time-Infeasible Project

Project BuyRite Case Study

Consider the following scenario for Project BuyRite towards the end of the planning phase.

> Paul Myer, as project manager, has completed a particularly thorough and reliable workplan (including a detailed WBS and well-constructed Gantt Chart). This reveals that (given assumptions about the project's budget and resources) it will take 30 months to produce all of the outputs that are specified in the scoping statement. In response, Nancy Palmer, declares that 30 months is too long and that she will allow only 24 months for the work.
>
> Paul Myer simply says "Oh, all right then" and commits to 24 months without providing any evidence of achievability. The project later gets into serious difficulty and is aborted.

Paul Myer could then face a charge of professional negligence because, he not only agreed to an infeasible project, he also failed to point out that his analysis has revealed that the desired timeframe cannot be met.

Situations of this kind can arise not only with timeframes, but also with budgets. Box 6.9 discusses the case of an infeasible budget. Satisfactory resolution of the problem requires not only a deep understanding about what is going on, but also knowledge of some simple analytical tools.

Firstly, both Paul Myer and Nancy Palmer have to be absolutely clear about the processes in which they are each engaged when they discover that their timeframes are incompatible. If Paul Myer has strong evidence of achievability for his 30 month figure, then he is involved in a process of estimation. The dates to emerge from time-estimation processes are called milestones. If Nancy Palmer's 24 month figure is actually "an opening bid", then we can conclude that she is involved in a process of negotiation. The dates to emerge from negotiation are called deadlines. Milestones are derived from an analytical model of the project's schedule, while deadlines are arbitrary dates without analytical support. Projects can be managed to milestones, but they cannot be managed to deadlines. Both estimation and negotiation are valid processes, but as long as they produce incompatible dates, they are in conflict.

To reconcile the two processes, the first point that both Paul Myer and Nancy Palmer have to recognise is that given the resourcing assumptions used in the estimate (of 30 months), the 24 month timeframe (with its attendant deadline) is infeasible. (Infeasibility is revealed by the fact that on a Gantt chart, the end-of-project milestone lies to the right of the desired deadline). The project *must not proceed* until this infeasibility is resolved. The infeasibility will have been resolved when an acceptable deadline lies on, or to the right of, the end-of-project milestone. Only three strategies are available to address this time infeasibility, which Paul Myer must now explore to see if any mix of them will satisfactorily resolve the situation:

1. Apply extra resources (to selected tasks).
2. De-scope the project.
3. Relax the deadline.

We now examine each of these in turn, and decide on their implications for the overall attractiveness of the project.

The application of extra resources inevitably involves higher costs, such as overtime or premium rates, with the result that reduced time usually means extra cost and hence lower worth. That is a trade-off that only the project owner can make (in consultation with the funder). While it is the responsibility of the project manager to identify and cost the options, he/she cannot make a decision that impacts the worth of the funder's investment. Whenever the "extra resource" option is exercised, its effects can be interpreted as a reduction in time frame brought about by increasing project cost (although, in reality the two are only indirectly related).

Descoping the project involves either removing outputs, or lowering the fitness-for-purpose of selected existing outputs. While such a move reduces both the timeframe and cost, it will also reduce the project's benefits (because utilisation is compromised). In general, these descoping effects, when all taken together, lower the worth of the project. (If that were not true we would have to conclude that the project had been poorly scoped to begin with). Again such an option can only be exercised by the project owner in consultation with the funder.

Finally, we have a third option, that of relaxing the deadline, which can have no impact on the project's:

- Timeframe (which is defined by the underlying milestones, not by desired deadlines)
- Cost (as distinct from its allocated budget)
- Target outcomes

and so the project's estimated worth is left intact. Despite this (highly desirable) state-of-affairs, the stakeholder who proposed the deadline must, of course, agree to its relaxation.

In general, resolution of a time infeasibility will involve a combination of all three strategies.

However, because 'one size does not fit all projects' (Shenhar, 2001), the list provided in Table 6.1 cannot be used as it stands without reference to the differences among projects, as might be found, for example, across industries (Cooke-Davies & Arzymanow, 2002; Ibbs & Kwak, 2000; Pennypacker & Grant, 2003). Such a view may indicate that different industries require peculiar sets of critical

Table 6.1 Ranked planning processes

Ranking	Planning process	PMBOK's knowledge area
1	Staff acquisition	Human resource
2	Project plan development	Integration
3	Cost estimating	Cost
4	Activity definition	Scope
5	Risk management planning	Risk
6	Quality planning	Quality
7	Resource planning	Human resource
8	Procurement planning	Procurement
9	Schedule development	Scheduling
10	Communications planning	Communications
11	Activity duration estimating	Scheduling
12	Scope definition	Scope
13	Cost budgeting	Cost
14	Activity sequencing	Scheduling
15	Scope planning	Scope
16	Organisational planning	Human resource

success processes. As a result, key players should be aware of the CSPs that apply to particular project scenarios so that they can ensure that they receive appropriate attention, according to the context of the project in hand. While the full results are presented in Appendix C, the most critical planning process for a selection of industries is presented in Table 6.2. All these results were found to be statistically significant.

Just as differences emerge across industry sectors, based on cultural diversity theories (Hofstede, 2001; House, Javidan, Hanges, & Dorfman, 2002) different results are experienced across countries. For example, in the research study (Appendix C) discussed earlier it has been found that staff acquisition is the most critical planning process in Japan, quality planning in New Zealand and activity definition in Israel. Discussion of cultural differences in project management among these countries can be found in Zwikael, Shimizu, and Globerson (2005) and Zwikael (2009).

Finally, different CSPs have different impacts on various project management success dimensions. Appendix C shows how critical planning processes (grouped by various success measures) differ with project focus. Table 6.3 summarises the results discussed in Appendix C by highlighting the most critical planning processes for four types of project focus.

Although this section highlights the need to give special attention to the planning processes that are most relevant to particular project scenarios, Zwikael and Globerson (2006) found that project managers usually do not divide their time appropriately. Generally speaking, they tend to spend more time on planning processes of a technical nature, since they are easy to perform, regardless of the extent to which such processes contribute to project success. However, these easier-to-do processes are often carried out at the expense of others that are much more important. For example, the actual time devoted to "communications

Table 6.2 Highest ranked critical planning processes for various industries

Industry	Highest ranking CSP in the industry	Conjectured reason
Engineering	Project plan development	The high levels of complexity of projects in this sector demand correspondingly high levels of detail in the plans used to guide them
Software and communications	Cost estimation	Given the high levels of uncertainty of estimates in this dynamic industry, a focus on estimation can go a long way towards improving the reliability of the plan
Production	Staff acquisition	Most employees in this industry are highly experienced with operational processes, and so care should be taken in choosing people who can cope with the project environment
Construction	Risk planning	This industry traditionally deals with generic risks that are readily identified and mitigated (bad weather and rejected permits are typical examples), hence mitigation planning is important
Services	Schedule development	Because this industry has relatively low levels of project management experience, basic tools and techniques are essential
Government	Activity definition	A heavy reliance on standing structures makes it extremely difficult for many government organisations to accommodate the temporary governance arrangements demanded by projects. Accordingly, it is important to identify all those who should be involved in a project and ensure that tasks are assigned carefully and formally

Table 6.3 Critical planning processes for different areas of project focus

Success test	Project focus	Critical successes processes during planning
Project management success	Timeframe	Schedule development
		Quality planning
		Communication planning
	Cost	Schedule development
		Cost estimation
	Scope/quality	Staff acquisition
		Activity definition
		Project plan development
Project ownership success	Funder satisfaction	Staff acquisition
		Project plan development

planning" and "quality planning" is less than would be expected, (based on their contributions to project success). The relative neglect of these processes may reflect factors such as the quality of available tools and the levels of competence in these particular skills amongst project managers.

6.2.3 The Work Breakdown Structure

Once all project outputs have been confirmed in the approved business case, then attention can turn to the process of producing them. The work breakdown structure (WBS) is a hierarchical model of all the work required to produce all of the project's committed outputs. The WBS for the project overall is made up of the WBSs for each of the project's outputs. Zwikael and Globerson (2006) found that the use of the WBS for reliable identification of a project's tasks is one of the most critical processes undertaken during planning.

Depending on the complexity and size of the output, this hierarchy may involve many levels. For ease of description in what follows, we have adopted a common stylised three-level hierarchy which is assembled by considering each output in turn.

- Level 0: *The output name*.
- Level 1: *Phases*. At its highest level, the work to build an output is described with a sequence of *about* seven large-scale steps, called "Phases". Take an office relocation project for example, with one output called "a moved office". The work involved here might be broken out into the following phases: Determine requirements for new office, Find new office, Design layout, Fit out, Move in, Resume regular business.
- Level 2: *Activities*. Take each "Phase" and break it out into *roughly* seven steps, called "Activities". In the office relocation example, "Find new office" could be defined by the following activities: Commission premises consultant, Locate suitable office spaces, Evaluate and rank options, Negotiate lease for preferred location, Finalise lease.
- Level 3: *Tasks*. Take each Activity and break it out into (again *approximately* seven) steps, called Tasks. A task is the smallest bundle of assignable work that will be recognised during the execution of the project. Tasks become the "work packages" that are assigned to team members.

This simple three-level structure is suitable for small to medium projects, but inadequate for larger exercises, which require more levels in the work hierarchy. Clearly in such cases there is little point to giving the levels names (such as Phase, Activity and Task) and so everything is simply called a "task" or an "activity". (Readers who use scheduling products such as Microsoft Project, for example, will be familiar with this convention).

The highest level of a WBS for a specific class of output is sometimes (and perhaps rather confusingly) called a "methodology". A case in point is the waterfall approach commonly used in the Information Technology industry which recognises generic phases for software, such as: analysis, design, development, testing and implementation, as the foundation of a "software development methodology".

The WBS serves as a "catalogue" of the steps that must be taken to produce all of the project's outputs. This now allows, using a "bottom-up" approach (discussed in the next section), validation of a number of critical parameters for which values had originally been set in the business case (especially those related to timeframe and resources).

In effect, a WBS replaces a single (large) process with a large number of small tasks. For example, in project BuyRite, the process "Produce a panel of preferred suppliers" is represented by a number of small tasks (such as "Select panel members"). The procedure of taking a small number of large things (many of which are too big to understand in a meaningful way) and systematically replacing them with a large number of small things (each of which is easily understood) is called "hierarchical decomposition". Hierarchical decomposition is a completely general analytical technique that is used in many disciplines. (Perhaps the best known illustration is Linnaeus' taxonomy of all organisms).

6.2.4 The Gantt Chart

The detailed model represented by the WBS can now be employed to estimate the duration of the work involved in the production of all the project's outputs. This is obtained by estimating the duration of each task and noting any dependencies amongst those tasks. The process is called scheduling, the output from which is a timetable of some form. The Gantt chart is a common and effective way to present a project's schedule in which a horizontal bar, set against a time scale, indicates the start, duration and end of each task in the WBS.

Scheduling allows a number of important questions to be answered, especially concerning any deadlines that may have been proposed:

- "Is the deadline feasible?"
- "If not, what must be done to make it feasible?"

The following major steps are performed to develop a Gantt chart:

1. *Estimate duration*. The elapsed time (for example, in days) is estimated for each task in the WBS. As explained in Box 6.5, the duration of a task is determined effectively by the allocation of resources.
2. *Define dependencies among tasks*. For example, in an office move project, the task "Unload the vans" cannot start until "Drive vans to new location" is finished. The effect of this is to constrain certain tasks (by setting the earliest dates on which they can start).
3. *Develop Gantt chart*. There are two broad approaches to deciding the feasibility of a deadline: forward scheduling and backward scheduling. In forward scheduling, we begin by considering the start date of the project and work forwards to calculate its end date, based on the durations of tasks/activities and dependencies among them. This procedure eventually reveals the *earliest* date by which the last task in the project can be completed. The feasibility of any proposed project deadline is readily decided by noting whether or not it occurs

Box 6.5 Illustration of Concept: Work Breakdown Structure (WBS)

Project BuyRite Case Study

One of the outputs proposed for Project BuyRite (the ICO case) is *a panel of preferred suppliers*. When assembled, this would take the form of a list of organisations from whom, under stated conditions, business units must purchase supplies for ICO.

The work involved in creating such a panel could be described using three-level WBS structure discussed above. For example, the highest level could be defined with six phases:

1. Analyse current supplier arrangements.
2. Design framework for pilot panel
3. Assemble pilot panel
4. Conduct pilot test and evaluate
5. Design operational panel
6. Implement operational panel

Phase #3 (*Assemble pilot panel*), could then be broken into four Activities:

3.1. Define supplies to be covered by pilot
3.2. Confirm membership of pilot panel
3.3. Induct members into pilot panel
3.4. Implement pilot

Similarly, activity #3.2 (*Confirm membership of pilot panel*), could then be broken into seven tasks:

3.2.1. Identify candidate panel members
3.2.2. Revise selection criteria
3.2.3. Survey/interview candidate panel members
3.2.4. Rank candidates
3.2.5. Invite organisations to join pilot panel
3.2.6. Select panel members
3.2.7. Formalise appointment of panel members

Each of these third level tasks is then delegated to a team member. Notice the word structures used for the items in the above lists, they all take the form of *imperatives* (commands). An imperative is how an instruction is expressed for someone to complete an action of some kind. A principle applied to the construction of WBSs is that activities and tasks should be expressed as imperatives. There are exceptions to this rule, but it provides a powerful way of ensuring that work and the outputs from that work are not confused.

The power of a WBS derives from two simple characteristics: it is hierarchical, with each node representing "sevenish" (approximately seven) items. Take the seven tasks given as examples above. It would be relatively easy for an expert to confirm that no task had been omitted (and that no redundant tasks had been included) because the list is equivalent to the activity "*Confirm membership of pilot panel*". Similarly, the list of activities (which includes "*Confirm membership of pilot panel*") is equivalent to the Phase "*Assemble pilot panel*". Therefore, when assembling or validating a WBS one is only ever concerned with examining a structure made up of lists of "sevenish" items. It may take a long time to develop and confirm a large WBS, but the work involved is relatively straight forward.

Why *sevenish*? Research by psychologists over many years (Miller, 1956) suggests that humans are quite good at manipulating about seven chunks of information. Reducing the count of components into which each item is broken increases the count of levels in the hierarchy (making it more complex). Increasing this number (generating potentially long lists of components), reduces the count of levels in the hierarchy (making it simpler), but then each list becomes difficult to handle.

after the estimated finish date of the last task. In backward scheduling, we begin by considering a desired deadline as the project end date and working backwards to the start. Again the durations of each task are estimated based on the assumed availability of resources. Dependencies are now used to establish the dates by which each task must be *started* (from the last, back to the first). This procedure eventually reveals the *latest* date by which the *first* project task must be started if the desired end date is to be met. The feasibility of the deadline is readily decided by noting whether this latest start date occurs after work on the project can start.

4. *Assemble a Schedule of Milestones*. A schedule of milestones is assembled so that the eventual progress of the project can be tracked. This takes the form of a table or chart that shows the planned completion dates of particularly noteworthy tasks (such as those associated with delivery of a major output or end of a phase). Because this chart shows *what* should be done (rather than *how* it is be done), some scholars have found that it promotes result oriented thinking (Andersen, Grude, & Haug, 1995; Turner 2009).

The schedule of milestones is abstracted from the WBS and so two questions need to be answered:

– How many milestones are required?
– How are they selected?
  Because milestones are used for tracking, there should be as many entries in the tables as there are review points for the project. In general, there should

Box 6.6 Practicalities

The Two Fundamental Approaches to Estimation

The core parameters covered by a project plan involve estimates of a project's duration, outlays, labour (and possibly other specific resources). It should be noted that there are two broad approaches to estimating these parameters: bottom-up and top-down.

Bottom-up estimation, also called work-based estimation, involves analysing each task (at the "bottom" of the WBS) in terms of: labour, outlays and duration, and then aggregating these values for the project overall. In this approach, we ask three questions about each task (the lowest level of the WBS):

1. *Outlay*. What do we need to purchase for this task, and how much money must we outlay for that purchase?
2. *Labour*. Who needs to be involved in this task, and how much labour (in person hours) is required of them?
3. *Duration*. Given resource assumptions and availability, how long will it take to complete?

Of these three parameters, the first two can be simply summed to get the corresponding figures for the project overall, but the third requires the application of some mathematical techniques (typically provided in project scheduling software).

Take, for example, the task of training staff over 5 days in use of a new operational process. Assume that a training centre must be hired at $4,000 per day, and that two internal specialists from the organisation's Learning & Development Department will run the course. In this case the task requires an outlay of $20,000 and 10 person-days of internal labour.

Bottom-up estimation is not always appropriate, especially when reliable estimates can be attached more readily to *outputs* than to the *work of producing those outputs*, as is often found in construction and engineering. In that case a top-down approach appears to be extremely effective.

Top-down estimation also called output-based estimation, involves analysing each output and, based on the experience with similar outputs produced in past projects, directly generating estimates of labour, outlays and duration.

Consider a project to build a wharf for a ship-loader that requires 250 lineal metres of sheet piling. Outlays on material and labour, as well as the time involved would be based on unit coefficients derived from past projects (adjusted for the peculiarities of the subject project). If, for example it was found that costs and productivity ran at about $2,500/lineal metre and 7.5 lineal metre/day respectively, then a top-down estimate could be derived quite readily.

Because, in most projects the original business case includes only a high-level WBS, the estimates of labour, outlays and duration provided there will have been derived from a top-down approach. If it is believed that bottom-up estimates will be more reliable than top-down figures, then the bottom-up approach will allow the figures provided in the original business case to be validated.

It should also be noted in passing that a formidable range of estimation tools has been assembled over the past two or three decades. Unfortunately, discussion of these lies beyond the scope of this book.

be as many milestones as there are project review points. If, for example, a 12-month project will be subject to weekly reviews by the team, then the schedule of milestones will have about 50 entries. These need not be equally spaced, but the intention is not to have lengthy periods when no checkpoint on progress is available. As far as the selection of tasks to use as milestones goes, three rules of thumb are suggested:

– they should be "noteworthy" or "significant".
– they should be scattered to occur approximately every week.
– they should all be on the critical path. Other milestones may also derive from contract commitments, for example payment times, formal reviews and deadlines mentioned in the project contract.

An illustrative sample of milestones for Project BuyRite is provided in Box 6.7.

5. *Identify the critical path.* The critical path is a chain of linked tasks that, if not completed on time, will necessarily delay the completion date of a project (by the same amount). The critical path can be easily identified using most common project scheduling software packages. In addition to milestones, other activities on the critical path should be monitored closely by the project manager during execution, as any delay with these will cause the whole project to finish later than planned.

6. *Identify activities with large slack.* Slack is the amount by which the duration of an activity can be extended, without causing the project to be delayed. As these figures are easily identified using a scheduling software package, project managers can use activities with large slack to increase their flexibility in planning. For example, resources from such activities may be moved to activities on the critical path, if the result is shorter overall project duration.

7. *Address timeframe issues in the business case.* Planning may reveal issues with the timeframe established in the approved business case. For example, it may now become clear that, given constraints on scope and budget, the original timeframe cannot be achieved. Even if the original timeframe was feasible, pressure may emerge during planning for an unrealistically short project. Before finalising a plan, project managers (in consultation with the project owner) are required to replace all deadlines with milestones. More is said of this in Box 6.7.

Over the past few years, other approaches have been proposed to deal with the phenomenon of budget (and time) overruns, such as the "Agile" development methodologies of the information system sector. These are based on the view that only time and schedule constraints should be set for a project. In that case, scope is determined by whatever has been delivered when the first of the constraints is struck in the current cycle (Anderson, 2004). In Chap. 5, we propose a more general approach to managing extremely high levels of uncertainty in projects, also based on iteration.

6.2.5 The Project's Estimated Cost

The estimated cost of a project should be distinguished from its budget. A project's budget is defined as the pool of money approved to cover project outlays. A project's cost is defined as the outlays required to purchase resources for the project. The term "budget" is frequently used to mean an arbitrary pool of funds that has been allocated to the project (often well before a business case is prepared), but for which there is no supporting analysis of resources. Because this use of "budget" is inconsistent with the definition used here, it can cause confusion, or even lead to a "cost-infeasible" project (See Box 6.9). Take, for example, the case of an organisation that has a history of budget "overruns" (because of unachievable budgets imposed on past business cases). In an attempt to pre-empt such a situation from arising on a new project, the funder may well set an artificially low budget as an (unstated) ambit claim, justified as being a "constraint". Because in that case, the budget has no supporting analysis and "evidence of achievability" it cannot be used as a form of estimate in the business case.

The issue of whether or not to separate cash outlays from the notional value of internal labour involves some concepts that we do not cover here. For the moment, the immediate discussion applies equally to both purchased-in resources (for which real money will be required to make the necessary marginal outlays), and to internal labour (for which there will be some sort of notional headcount allocation or assignment of specific staff). Regardless of the treatment of labour, the project manager has a responsibility for producing reliable, verifiable estimates of cost, on which the project owner (in consultation with the funder) can then base a budget.

Section 3.3 introduced a three-way classification of project costs: production, management and (eventual) operations. The project budget is based on the first two of these (which relate to above-the-line and below-the-line activity respectively). The third is required for the financial analysis of the project. Planning will normally result in more reliable estimates of all three, and provide the foundation for the financial appraisal section of the modified business case and the schedule of outlays in the project plan. Cost estimates are refined during planning and are highly dependent on the project schedule. In general, costs can be reduced by extending the project's duration (or, equivalently, the project's duration can be reduced by selectively accepting extra cost). The relationship between project

Box 6.7 An Explanation of Underlying Principles

How are Estimates of Durations for Tasks Derived?

Experienced estimators establish durations for tasks based on some notional allocation of resources. If, at any time it becomes apparent that the assumed level of resource allocation is wrong, then they will adjust their estimates accordingly. The principles that underlie this practice are revealing, but, unfortunately, not always understood.

The duration of a task is dependent only on:

- The "intrinsic load" represented by the task in question. For example, the intrinsic load represented by a proposed task in Project BuyRite "survey 15 suppliers" is clearly less than would be expected from the task "*survey 30 suppliers*".
- The notional allocation of resources to the task, for example, *three interviewers*.
- The proficiency of each resource. For example, each interviewer may be capable of interviewing *two suppliers per day*.

These variables can be related in a simple equation;

$$Duration = Load/(Resources \times proficiency)$$

For example, surveying 30 suppliers using three interviewers (with the stated level of proficiency) would, therefore, take 5 days ($30/(3 \times 2) = 5$).

In most real-life situations, the load from each task is given, as is the proficiency of available resources, and so the only discretionary variable normally available to the estimator which he/she can use to alter task durations is the level of resourcing.

Notice how in forward scheduling, *finish dates* are dependent on durations and in backward scheduling, *start dates* are dependent on durations. Durations must *never* be calculated as the difference between two set dates. Inexperienced project managers, when faced with deadlines (especially if imposed by senior managers), automatically assume that their analysis is, in some sense "wrong". In an attempt to "correct" the "error", they will go back through their work and change all of their estimates until the timeframe is now consistent with the imposed deadline. If the original *estimated* timeframe was accurate, any adjustments of that kind, will of course, lower the reliability of the project plan, they cannot possibly improve its quality, and so are professionally unacceptable.

The achievability of a timeframe is decided by the *quality of the schedule*, not by the *urgency of deadlines*, nor by the *level of optimism* reflected in the schedule.

A similar situation can arise with budgets. If Paul Myer had produced a highly reliable estimate of project outlays, only to be told that "He could be given only half that amount", he would again be required to point out that such a budget was inadequate and that the project should not proceed until the issue had been resolved.

The reader should note that all estimates are subject to uncertainty and that the reliability of timeframes must be declared in the relevant baseline document.

Box 6.8 Illustration of Concepts: Gantt Chart and Schedule of Milestones

Project BuyRite Case Study

Consider the activity numbered #3.2 ("Confirm membership of pilot panel") in the previous Project BuyRite Case Study box (entitled "Illustration of concept: Work Breakdown Structure"). Assume that, with available resources, the duration of each component task has been estimated as follows (Table 6.4):

Table 6.4 Project scheduling input for a part of Project BuyRite

#	Task name	Duration (weeks)	Immediate predecessor
3.2.1	Identify candidate panel members	1	–
3.2.2	Revise selection criteria	1	–
3.2.3	Interview candidate panel members	2	–
3.2.4	Rank candidate panel members	1	3.2.3
3.2.5	Invite organisations to join pilot panel	2	3.2.2, 3.2.3
3.2.6	Select panel members	2	3.2.4, 3.2.5
3.2.7	Formalise appointment of panel members	2	3.2.6

Some of these tasks are serially dependent, while others can be done in parallel and so a Gantt chart for this portion of the WBS for Project BuyRite might look something like this (Fig. 6.3):

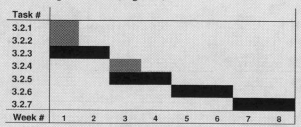

Fig. 6.3 A part of the Gantt chart for Project BuyRite

Note how the table indicates clearly that "Rank candidate panel members" (#3.2.4) cannot start until "Interview candidate panel members" (#3.2.3) is complete, and that selection of panel members (#3.2.6) can be started only after *panel members have been ranked* (#3.2.4) and *invitations have been sent to organisations* (#3.2.5).

The critical path of this project (highlighted in the Gantt chart with darker bar colour) will include four activities that are to be executed one after the other: #3.2.3, #3.2.5, #3.2.6 and #3.2.7. The project manager will have to monitor the progress of these activities closely, as delay with any of these will immediately cause the following critical activity to start late and, as a result delay the entire project.

If it is anticipated that Project BuyRite will run for 30 months and it is intended that the project team will meet weekly, then the eventual schedule of milestones will have about 75 entries. (The WBS for the project will, of course have many more entries than this). It would be reasonable for Paul Myer to constrain selection of these milestones, by imposing rules such as: there are never more than two in one week and there is never more than one week without a milestone.

The phases suggested in Box 6.4 for the output "Panel of preferred suppliers", could well be included in the eventual schedule of milestones for the project overall, thus providing six of the nominal 75 entries:

Analyse current supplier arrangements	25-Mar
Design framework for pilot panel	20-Apr
Assemble pilot panel	16-Jun
Conduct pilot test and evaluate	30-Aug
Design operational panel	25-Sep
Implement operational panel	3-Dec

duration and cost appears somewhat counter-intuitive and so the matter bears some further discussion. In order to do that we have to consider the way that costs are driven by above-the-line and below-the-line activity.

Box 6.9 Illustration of Concept: A Cost-Infeasible Project

Project BuyRite Case Study

In Box 6.8 we explored the concept of a time-infeasible project. Here we consider a closely related problem, that of a cost-infeasible project, by hypothesising a slightly different scenario for Project BuyRite. Again, to allow for the two problems to be compared and contrasted we assume that the following situation has emerged towards the end of the planning phase.

Paul Myer's workplan reveals that (given assumptions about the project's timeframe and resourcing) the exercise will require a budget of $45M. (This figure accounts for all purchases of external resources and the opportunity cost of all internal staff). In response, Nancy Palmer, declares that $45M is too much and that she will approve no more than $35M for the project.

As in the previous case, if Paul Myer agrees to the reduced budget without providing any evidence of achievability and the project is eventually aborted, he could then face a charge of professional negligence (for similar reasons to those outlined in the time-infeasible case).

Again, both Paul Myer and Nancy Palmer have to distinguish between the processes of estimation and negotiation, this time with the focus on funds.

To reconcile the two processes, both players must recognise that given the resourcing assumptions used in the estimates for the timeframe (of 30 months) and outlays, the $35M budget is infeasible. (In other words the scope and timeframe are inconsistent with a $35M constraint on funds). Again, only three strategies are available to address this cost infeasibility, all of which Paul Myer must now explore to see if any mix of them will satisfactorily resolve the problem. One of the three strategies in this case also appears in the list related to the time-infeasible project. The other two are subtly different.

1. Increase the project timeframe.
2. De-scope the project.
3. Increase the allocated budget.

Similar to the discussion of time-infeasibility, we now consider each of these in turn, and decide on their implications for the overall attractiveness of the project.

Increasing the project timeframe allows low-price resources to be substituted for high-price resources, as discussed above, with the result that reduced cost means extra time. In this case, the impact on worth is more difficult to predict (because while lowered costs increases worth, delayed benefits lowers worth) and so a decision would have to be made on a project-by-project basis. Again, that is a trade-off that only the project owner can make (in consultation with the funder).

The implications of descoping the project have been covered earlier.

Finally, we note that increasing the allocated budget can have no impact on the project's:

- Cost (which is determined by the underlying resource assumptions)
- Timeframe
- Target outcomes

As was the case with relaxing an arbitrary deadline, the project's estimated worth is left intact. Again, the stakeholder who proposed the budget must, of course, agree to its relaxation.

In general, resolution of cost infeasibility will involve a combination of all three strategies.

Below-the-line cost arises from the work to produce below-the-line outputs. Below-the-line outputs are those that appear in the project's statement of scope. Below-the-line cost and project duration behave as if they are inversely related, but the relationship is quite subtle. Neither cost nor duration can be manipulated directly, but both are sensitive to changes in a third variable that can be manipulated by the project manager, related to the levels of resources applied to tasks in the WBS. If resources are withdrawn selectively from a task, in general its duration will increase and its cost will decrease. Consider, for example what happens if we eliminate some agreed overtime to complete a task. Instead of gaining quick access to a skill (by paying someone to work longer hours), we must now wait for access to that same skill as it becomes available in normal working hours. The overall effect of this strategy is to lower cost (because normal hours are cheaper that overtime hours), but slow the rate at which the work is completed.

If, on the other hand, additional resources are selectively applied to a task, by reverse reasoning, its duration will decrease and its cost will increase. These effects can be summarised as two "rules":

- Above a certain threshold, to reduce the duration of a task, selectively apply additional (high-price) resources. In general this will increase its cost and so it appears that duration is falling in response to a cost increase.
- Above a certain threshold, to reduce the cost of a task, selectively withdraw (high-price) resources. In general this will increase its duration and so it appears that cost is falling in response to an increase in duration.

In practice, numerous strategies are available to reduce duration, including the use of airfreight instead of surface shipping and working overtime (both of which are examples of "high-price" resources). Conversely, to reduce below-the-line costs, decisions might be taken to use surface shipping instead of airfreight and avoid overtime. The apparent relationship between below-the-line cost and duration—is shown in Fig. 6.4, where below-the-line costs are reduced by allowing duration to rise.

Above-the-line costs (associated with the planning, management and administration of the project) arise from above-the-line outputs (such as a monthly status report). These outputs are generic to most projects and often not included in the

Fig. 6.4 The effect of resource manipulation on total below-the-line cost and duration

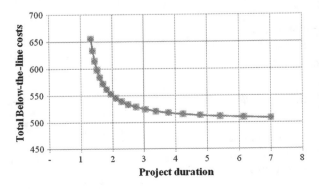

Fig. 6.5 Above the line costs responding to project duration

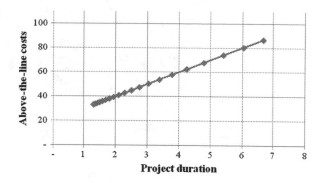

Fig. 6.6 The total cost of a project responding to duration

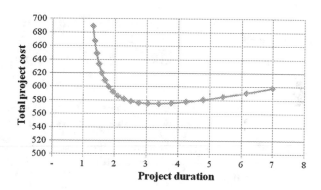

WBS (which is based on outputs included in the statement of scope). In some cases these costs also include overheads (such as office rental and the fees for a part time administrator). Because above-the-line costs arise from such activities as meetings, reports and the use of administrative tools, it is clear that they are driven primarily by the duration of the project (longer the project, the more meetings and reports that will be required). The behaviour of above-the-line costs in response to project duration is suggested in Fig. 6.5.

The total cost of a project is the sum of both types of costs, as is presented in Fig. 6.6. An appropriate choice of time/cost strategy will depend on the project's pressures and constraints at the time. For example, if total project costs have to be minimised, then duration may need to be varied (assuming no change in scope or risk exposure). If, however the duration of the project must be minimised, then increases in costs have to be accepted.

6.2.6 Risk Mitigation Planning

Before outlining a process to manage project risk, an analytical framework is needed on which to base the discussion. We use an event-impact model for risk, as shown in Fig. 6.7.

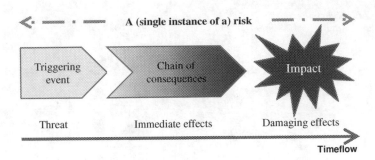

Fig. 6.7 The event-impact model of a project risk

The event-impact model is a mechanism involving three time-related components:

- A triggering event, which takes the form of a threat. For example "Project manager resigns".
- A chain of (immediate) consequences, for example, if the project manager leaves there will be no one to conclude the negotiations with the subcontractor.
- A (damaging) impact, which lowers the worth of the project.

We propose that the event-impact model is a completely general representation of risk, and, accordingly, that all forms of risk can be expressed as specific instances of the mechanism.

Since the multitude of threats that a project faces during its life vary widely in their significance, we need a method to make systematic judgements about the importance of each. This will enable risks to be ranked so that attention can be directed at those considered "important". To do this we define a measure of "importance" which is based on both the likelihood that a threat will emerge and the severity of the damage to the project's worth if that threat arises. This measure is called the project's "risk exposure".

Threats have a likelihood of occurrence which can be viewed as a (qualitative) level of belief or expectation by someone that the event will happen. Likelihood is the inverse of the "level of surprise" that someone would experience on discovering that the threat had occurred. Likelihood will be moderated by the information about the threat that is available. For example, if in the ICO case study, Nancy Palmer (the project owner) became aware that Paul Myer (the project manager) was very unhappy, then she may well form a view that his resignation was "highly likely". Likelihood is judged qualitatively, unlike probability, which can be estimated empirically from data. Accordingly, likelihood can be gauged either using word scales, or using a metric that (like probability) lies between zero and one.

The likelihood of a threat can be time-dependent. A prime contractor on a construction project may have a greater likelihood of declaring bankruptcy early in the project than later because of the nature of his cashflow. This implies that threats should be regularly reviewed and updated.

The chain of consequences (shown in the middle of Fig. 6.7) does not have to be explicitly described or articulated, it is simply a way of ensuring that a plausible cause-effect relationship has been established between the triggering event and the impact.

A damaging impact is gauged by its "severity", related to the total reduction in project worth if the threat were to be realised.

Given the likelihood of a threat and the severity of its damage to the project, we can determine the risk exposure of the threat. A procedure for doing this is described below.

Using the event-impact model, we are able to classify a number of common situations surrounding project risk, and draw conclusions from this about how such situations are to be managed. The first situation is where we can identify specific threats, such as "project manager leaves". Here we are able to complete the entire risk analysis process, and, if the risk exposure is high enough to be of concern, consider actions to mitigate it. The second is where we become aware of a damaging impact, but where the triggering events are unknown. For example, we might believe that our costs could rise as the end result of any of a large number of unknown causes. Here we may only be able to describe the probability of exceeding delays of a given magnitude (and hence very limited in our ability to take mitigating actions). The third is where there are a very large number of known low-likelihood events, all of which contribute to a similar collection of damaging impacts. Because there are too many threats to analyse individually, we are restricted to the same sort of approach that applies to the second situation.

Despite the generality of the event-impact model, there is no correspondingly general technique of analysis or approach to risk management, and so we are forced to adopt different approaches to different forms of risk. Two important forms of risk are distinguished by the nature of their triggering events, which are either discrete or continuous. Consider the threat "Prime contractor declares bankruptcy", which is a discrete-variable event (because it either happens or it doesn't). By contrast the threat "Rain falls" is an example of a continuous-variable event because rainfall can range from being merely irritating (a single light shower) through to catastrophic (causing severe flooding).

While techniques exist for handling both forms, the available approaches to continuous event risk are not particularly suited to the project environment because of their relative complexity. In contrast, not only is the common approach to discrete event risk simple, it can also be adapted to cope with continuous event risk. Having said that, the limitations of this approach must be borne in mind when using it to gauge the "riskiness" of a project.

Risk management is ongoing throughout the entire project, starting very early during the initiation phase, when threats to the project are identified and analysed. The process then involves the assembly and implementation of an appropriate programme of risk mitigation (Table 6.5).

A risk mitigation strategy is a made up of agreed actions that seek to reduce the project's risk exposure (by reducing the level of importance associated with particular risks), but it contains no details about those actions. For example, it might

Table 6.5 The risk management processes, based on discrete-event risk

Risk management process	Project phase	Outputs	Leader	Approved by
Identification	Initiation (ongoing)	A list of threats	Champion	Funder
Analysis	Initiation (ongoing)	Risk register	Project manager	Project owner
Mitigation planning	Planning (ongoing)	Risk mitigation plan	Project manager	Project owner
Mitigation implementation	Execution (ongoing)	Mitigated risks	Project manager	Project owner
Mitigation monitoring	Execution (ongoing)	Updated risk mitigation plan Updated risk register	Project manager	Project owner

be decided to formally appoint and train a deputy, as one of the contingencies for the threat "Project manager leaves". At this point there is no detail about the proposed training programme and so the next step in risk management assembles a detailed plan for the strategy, requiring Work Breakdown Structure (WBS), workplans and so on. The project manager normally administers the process, but execution of the plan itself will involve a wide range of key players, including members of the steering committee. Substantive work on implementing the mitigation plan takes place during execution. At this time, the project manager will also monitor the effectiveness of the risk mitigation plan and seek to control its operation.

In summary, risk management continues throughout the entire life of the project, based on the processes presented in Fig. 6.8.

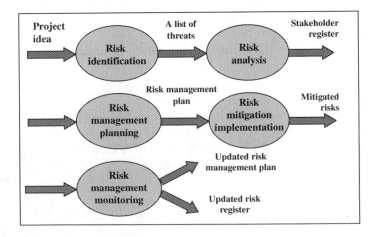

Fig. 6.8 Risk management processes

6.2.6.1 Risk Identification

Risk identification is concerned with the first element of the event-impact model. It is simply the process of identifying threats to the project. This is best done by small groups of people who understand relevant aspects of the project environment. The risk registers of past projects can be a valuable source of potential threats for new projects. We propose a particular word structure for threats, taking the form of "newspaper headline style statements", that is, they are expressed as they would appear on the billboards of a newspaper the day after they occurred. "Project manager leaves", "groundwater found to be polluted", "state government fails to pass enabling legislation", "new process proves slower than anticipated" and "prime contractor declares bankruptcy" are all examples of this format.

6.2.6.2 Risk Analysis

Risk analysis is concerned with assigning "values" for the likelihood of threats, working through the impact elements of the event-impact model and deciding on a set of actions that will be taken to mitigate each risk. The values assigned to the likelihood of a threat are drawn from some agreed scale, which can be numeric, or even a list of words. Numeric values for likelihood are often based on a number between 0 and 1 (like probabilities), or on wordscales, such as: high, medium, low. This second part of analysis involves three steps: identification of the particular forms of damaging impact that would arise from each threat, gauging the severity of that damage and "calculating" the resulting level of risk exposure. The third part of analysis is concerned with the selection of actions that will mitigate the risk and analysis of the effects of those actions on the project overall.

The damaging impact of a threat takes the form of a fall in the project's worth. Since worth is a function of benefits, disbenefits and costs, damage must be reflected in an adverse movement in one or more of these three variables. Furthermore, there are two ways in which each variable can move unfavourably:

- In magnitude, for example costs increased.
- In timing, for example benefits delayed.

Table 6.6 summarises the only possible six forms of damaging impact.

The severity of a damaging impact can be summarised as a (qualitative) measure of either a reduction in a project's worth, or as a delay in its realisation. The values assigned to the severity of a project's damage are drawn from some

Table 6.6 The six forms of damaging impact that a project can suffer as the result of a threat

	Magnitude	Timing
Benefits	Reduced	Delayed
Disbenefits	Increased	Advanced
Costs	Increased	Advanced

Table 6.7 A risk exposure grading table

Severity	Likelihood		
	High	Medium	Low
High	A	B	C
Medium	C	D	E
Low	E	F	G

agreed scale which (as in the case of likelihood) can be numeric, or a list of words. Selection of an appropriate value for severity involves two steps: the first is to identify which of the six forms of damage apply to the threat under analysis, the second is to make a judgment about the size of the associated damaging effects on worth. For the threat "the project manager leaves" in the ICO case, under normal conditions only two forms of damage would be expected from amongst the entries in Table 6.6: benefits delayed and costs increased. (Presumably, the impacts on worth would be similar).

At this point in the process, "values" have been decided for both likelihood and severity, thus allowing a "calculation" of risk exposure. Two common procedures for doing this use either a grading rule or a grading table.

A grading rule can be applied when numeric values have been assigned to both likelihood and severity. Assume that the likelihood of the threat "The project manager resigns" is judged as 0.6 (from a 0/1 scale) and that the severity of the damage if 80 (on a 0/100 scale). A grading rule such as "multiply likelihood and severity", would give a risk exposure of 0.6 * 80 = 48. Care must be exercised when interpreting numbers obtained in this way, for example, a risk with a risk exposure of 48 is not "twice" as important as another with a risk exposure of 24. We are justified in saying that it is "more important", but we cannot say by how much more.

A grading table achieves a similar result, but is used with wordscales. Consider an approach to the analysis of risk under which likelihood and severity are each drawn from a simple three-valued wordscale: high, medium, low. A grading table maps all combinations of the wordscale values for two parameters onto another (single) wordscale for risk exposure, as in the example shown as Table 6.7.

For the threat "The project manager leaves", assume that likelihood is judged as medium and severity as high, then the risk exposure is found to be "B" (in a wordscale that has the possible values of: A, B, C, D, E, F & G).

Grading tables based on non-numeric entries are not as susceptible to misinterpretation as numeric grading rules.

6.2.6.3 Risk Mitigation

A risk mitigation strategy is a selection of cost-effective actions that make a project "less risky" by reducing its risk exposure. There are two sorts of mitigating action that can be brought to bear against a threat:

- Preemptives that reduce the likelihood of a threat emerging.
- Contingencies that reduce the damage from any threats that do occur.

For the threat "project manager leaves", "pay a completion bonus" is an example of a pre-emptive, while "train a nominated deputy" is an example of a contingency.

By cost-effective we mean that the reduction in risk exposure from the action is adequate compensation for the costs incurred by taking that action. To make a considered judgement about a proposed risk mitigation action we need to know three pieces of information:

1. *Pre-risk exposure*. The risk exposure from the threat without the proposed action.
2. *Post-risk exposure*. The risk exposure from the threat if the proposed action is taken.
3. *The cost of the action*. The additional cost required to reduce the level of exposure from pre to post risk.

This, in turn, implies that mitigation planning requires a number of critical items of information about a risk.

6.2.6.4 The Risk Register

The primary tool in risk management is the risk register, which is used to document the results of analysis and outline the mitigation programme being proposed for each risk. The risk register is held in the business case and later in the project plan. Because it is a register, this tool takes the form of a table where rows are associated with instances (risks) and columns with attributes. A set of attributes making up a typical risk register would include columns for:

- *Threat*: description of the triggering event.
- *Pre-likelihood of the threat* in the absence of the proposed mitigating action.
- *The damaging impact on the project from the threat* expressed as one or other (or both) of two overall effects: worth reduced and worth delayed.
- *Pre-severity of this impact* in the absence of the proposed mitigating action.
- *The pre-risk exposure of the threat* in the absence of the proposed mitigating action.
- *The mitigation plan* a list of preemptives and contingencies that would mitigate the original threat.
- *Post-likelihood of the threat* assuming the proposed mitigating plan is put into effect.
- *Post- severity of the damaging impact* assuming the proposed mitigating plan is put into effect.
- *The post-risk exposure of the threat* assuming the proposed mitigating plan is put into effect.

- *The effectiveness of the proposed mitigating action* the difference between the pre and post risk exposures.
- *The cost of the proposed mitigating action* to within an order of magnitude.

Based on this analysis a decision can be taken about which mitigating actions merit inclusion in the project's risk mitigation strategy. In certain cases risk mitigation involves the production of additional significant outputs, beyond those identified in the early versions of the scoping statement. In that case the project scope must be expanded to accommodate these. Regardless of whether or not mitigation requires such deliverables, it certainly involves extra work, which the project's budget must accommodate.

The project manager holds the primary risk register, but (as was noted for the stakeholder register) there may be occasions when others hold "supplementary" registers. For example, the project owner may assemble a confidential risk register when premature release of information about a threat related to the project manager would cause irreparable damage.

6.3 The Project Team

During planning, the bulk of the load will be taken up by the project manager, using whatever resources that have been assigned (or that he/she can co-opt) to support the planning exercise. This planning team (if it exists) is relatively small, because the larger project team will not be formally created until the plan is approved. As a result, any group involved with the project manager in planning is usually quite ad hoc in nature, with much of the work being done in workshops, meetings and interviews by people who may not have been formally appointed to fill a planning role.

Planning teams are used for a variety of activities, but estimation is of particular importance. Based on their experience and knowledge those involved in this work assist the project manager to assemble the WBS, the schedule, estimates of time and cost for various project tasks and an assessment of the reliability of all estimates. The discussion about the reliability of estimates assembled during initiation (See Chap. 5) applies equally to planning.

Box 6.10 From the Literature

The Effectiveness of Risk Management

Both the business and project environments involve risk. The risks faced by project managers arise from a number of sources: technological (Ricci et al., 2002), financial (Hainaut & Devolder, 2007;Pongsakdi et al., 2006), insurance-related (Wattman & Jones, 2007), environmental safety (Qio et al., 2001), and venture capital (Smolarski et al., 2005). As a result, risk management is a crucial element in many business areas (Das & Teng, 1998; Kerzner, 2009; Kleindorfer & Saad, 2005; Kulkarni et al., 2004; Slattery & Ganster, 2002; Sodhi, 2005 ; Wallace et al., 2004).

Risk figures prominently as an element of project management (for example, Ford & Randolph, 1992). Huchzermeier and Loch (2001) identify five types of uncertainty in projects: market payoff, project budget, product performance, market requirements and project schedule. Project risk is a "measure of the probability and consequence of not achieving a defined project goal" (Kerzner, 2009). Because risk in projects cannot be completely eliminated, Chapman and Ward, (2004) have defined 'risk efficiency' as that form of risk mitigation which minimises risk to achieve some threshold of performance. Risk management deals with minimising risk levels by identifying and ranking potential risk events, developing a response plan and monitoring during project execution (PMI, 2008).

Consequently, in the literature, risk management is considered to be a critical area of project practice (Crawford, Pollack & England, 2006; Morris, Jamieson, & Shepherd, 2006; PMI 2008), with a variety of new tools being introduced (Ahmed et al., 2007). As risk management is highly developed in project management, most organisations have a formal policy for project risk management (Voetsch, 2004) and supportive tools (Ahmed et al., 2007; Kerzner, 2009). These tools include risk identification tools (e.g. brainstorming, checklists, influence diagrams, cause and effect diagrams), risk analysis tools (e.g. probability and impact grids, event tree analysis, sensitivity analysis and simulation, Delphi techniques, expert judgment), and risk evaluation tools (e.g. decision tree analysis, portfolio management and multiple criteria decision-making tools) and software packages (Herroelen, 2005).

Because of its importance, one would expect project risk management practices to be not only conceptually advanced, but also highly effective. However, some recent studies have raised a concern regarding the performance of risk management tools in the project environment (e.g. Fortune & White, 2006; Raz et al., 2002). Some shortcomings in current risk management practices, which have been identified in the literature in recent years include:

1. *Narrow focus.* Despite the wide variety of available risk management tools, most project managers use only a limited number, mainly concerned with ranking potential risk events (Chapman & Ward, 2004; Gray & Larson, 2006; Wallace et al., 2004).
2. *Poor application.* Project managers perform some of the most important risk management processes poorly, for example, risk identification (Kwak & Stoddard, 2004), and development of effective mitigation strategies (Wallace et al., 2004).

3. *Practicality.* As the size and complexity of projects increase, the effort required for effective risk management rises exponentially, making current tools very difficult to use (Kwak & Stoddard, 2004).

4. *Mismatch of authority and responsibility.* While project managers are normally responsible for risk management processes, it is often functional managers who have the necessary information and authority to effect them (Globerson & Zwikael, 2002).

5. *Low impact.* Risk management is ranked relatively low in studies identifying 'project critical success factors' (Fortune & White 2006). Worryingly, there is no strong empirical evidence in the literature that risk management has a significant impact on project success. Studies that encourage use of risk management techniques appear to suffer from some weakness, such as a self-selected sample of members from 'risk special interest groups', who do not represent the larger project management community (Voetsch, 2004).

6.4 The Project Owner

A project owner will normally have been appointed (as the funder's agent) when the business case is accepted. Because he/she is held accountable by the funder for the eventual realisation of the business case and hence its target outcomes, planning is notionally driven by the project owner, but led by the project manager. Despite the prominent role played by the project manager in planning, the funder will view the project owner as accountable for both the modified business case and the project plan. (In reality, the project owner will delegate accountability for the plan to the project manager). Accordingly, the project owner should seek regular briefings from the project manager so that he/she is completely comfortable with the final plan. The project owner may also find it appropriate to consult with the funder at various points in the planning phase.

6.4.1 Identifying Critical Outputs for Close Attention

While a project owner does not become involved with the same level of detail that is of concern to the project manager, he/she should be aware of areas of a plan that demand particular attention. An obvious example is the risk register where interest is centred on threats with high levels of risk exposure. This principle also applies to scope. A project manager is accountable for delivering all outputs fit for purpose, but the project owner should be especially concerned with those outputs that are going to have the most significant impact on the project's success. So, the question arises 'on which outputs should the project owner (and the steering committee) focus their attention during the course of execution?

The technique is proposed here that provides a quantitative framework to rank outputs according to their contribution to outcome generation. Whilst acknowledging that everything in the scoping statement must be delivered fit-for-purpose (within the constraints of time, cost and detrimental outcomes), the project owner is particularly concerned with those to which project failure is most sensitive.

The approach requires the funder to prioritise outcomes. Outputs can then be ranked by importance using the prioritised outcomes through the utilisation map (introduced in Sect. 5.2). The approach proposed here analyses the contribution of various outputs to outcome generation, according to the following steps, demonstrated in Table 6.8.

1. *Identify project outcomes.* The list of outcomes appears in the statement of scope. Record these outcomes in the "target outcomes" row.
2. *Determine the importance of outcomes to the worth of the project.* This can be evaluated with an index indicating relative importance (for example on a scale 1 through 5), or as percentages that sum 100%. The values are recorded in the second row ("importance to the funder").

Box 6.11 Practicalities: Handling Uncertainty in Costs

Budgeting versus Estimation

Budgeting is the process of sizing and securing the pool of funds that will be made available to a project, based on reliable estimates. The processes of estimating and budgeting are independent (although clearly the feasibility of the project demands that the agreed budget exceed the project's overall estimated cost). A common confusion about the relationship between estimation and budgeting often causes high risk projects to appear riskless. (Interestingly this confusion is propagated by the popularist treatments of projects seen in the press and electronic media).

Take the example of a budget that has to be set as a fixed figure, for a project with a high degree of uncertainty in its costs. While the uncertainty in costs is usefully expressed as a range, any demand for a fixed budget must not result in the range of cost estimates being replaced with an arbitrary point-estimate because point estimates imply certainty (that is a riskless project). The demand by a funder for a fixed budget in no way effects the uncertainty of the project and so estimates must be provided as a range, regardless of budgeting requirements. It is a professional responsibility of the project manager not only to provide reliable information about costs, but also establish very clearly the level of certainty surrounding estimates of labour and outlays. The project budget is informed by estimates, estimates are not informed by budget limits.

Table 6.8 The use of the utilisation map to calculate the relative importance of outputs—Project BuyRite

Committed outputs	Target outcomes		
	Reduced costs of procurement	Reduced payment times to our suppliers	Importance of outputs (the higher the number the higher the importance is)
Importance to the funder (1–5)	5	2	
New procurement processes	9	9	$5 \times 9 + 2 \times 9 = 63$
New procurement policy and procedures manual	5	5	$5 \times 5 + 2 \times 5 = 35$
Enabling applications systems & new technical infrastructure	5	1	$5 \times 5 + 2 \times 1 = 27$
Programmes of accredited professional development staff	5	1	$5 \times 5 + 2 \times 1 = 27$
A panel of preferred suppliers	1	0	$5 \times 1 + 2 \times 0 = 5$

3. *Identify project outputs.* The list of outputs (available from the project's Statement of Scope) is displayed in the "Committed outputs" column (as is done in the utilisation map). The scoping statement has four types of entry: ITO outputs (utilised by customers to generate target outcomes), risk outputs (required to mitigate certain risks), stakeholder outputs (required to engage certain stakeholders) and mandatory outputs (required by regulation, policy or law). It is only the ITO outputs that are relevant to this analysis.

4. *Define relationships between outcomes and outputs.* This indicates the degree to which each output contributes to each outcome (through utilisation by all customers). An index is appropriate with values indicating the strength of the relationship (say 1 through 9). Negative values can be used to suggest that utilisation of certain outputs actually attenuates the generation of some outcomes. These values are recorded in the table (using the cells that connect an output and an outcome).

5. *Calculate the importance of each output.* The contribution of each output to every outcome is calculated as a factor of two numbers—the importance of the outcome and the relationship between the output and the outcome. Then, for each output, the contribution to all outcomes are summed up and written in the "importance of output" column. For example, the output called "new procurement process" contributes to the first outcome "Reduced costs of procurement" by 45 points (impact of 9 to an outcome that is worth 5 points) and to the second outcome by 18 points (impact of 9 to an outcome that is worth 2 points), with a total of 63 (45 + 18).

6. *Rank the outputs by their relative importance.* List the outputs according to the scores calculated in the previous step.

As a result of this exercise, we can conclude that the success of the project is most sensitive to the development of new procurement processes. Care must be exercised with this approach because the ranking of outputs in this way assumes that project success is determined by each output independently of the others. If a situation emerges where success depends on a number of outputs jointly, then all those interdependent outputs would have to receive equal attention. This analysis suggests that, during execution, the project owner should pay particular attention to progress on the new procurement process, because of its impact on project worth. It must be pointed out, however, that the ranked list of outputs can only be used to focus the attention of key players, it cannot be used to prioritise outputs if it proves necessary during execution to descope the project. Descoping requires a much more comprehensive analysis using the utilisation map.

6.4.2 Approving the Project Plan

The project owner normally reviews final versions of the project plan and determines its suitability for execution, which will result in one of the following decisions:

1. *Approve*. Set up begins and the project progresses to the execution phase. The project plan as it stands becomes the foundation for the eventual project, although the project owner may first want to highlight certain elements in the document, such as:

   – Critical success processes to focus on
   – Project control processes
   – Steering committee involvement in the project
   – The next control step

   Comments, thoughts, observations and directives related to these points would normally be included in a formal project approval document. In the case of a significant initiative, the announcement of the project may be made in some sort of kick-off event to which all key stakeholders would be invited.

2. *Abandon*. The project plan is rejected. This will usually happen when estimates of the project's worth fall significantly short of expectations set in the business case (due to changes in anticipated benefits, disbenefits and costs) or when the risks surrounding the project appear unacceptably high.
3. *Rework the project plan*. Often the project owner will require more information before the project plan can be approved. In this case, the project plan will have to be reworked. When the required changes are minor, there may be no need for another presentation, in which case the amended project plan would be approved when submitted, and the work on the project can start immediately.

The project plan's acceptance by the project owner concludes the planning phase and project execution may begin.

6.5 The Steering Committee

The steering committee may be assembled at the beginning of planning or it may be delayed until set up (which is addressed after the project plan is approved)—this depends on the project context. At the very least there will be a "steering committee" of one (being the project owner) for the duration of planning. The decision on when to establish a steering committee is made by the project owner (perhaps in consultation with the funder). A steering committee may be useful during planning if the work involved is complex—as would the case when input was required from other organisational units (or other organisations). Any steering committee established for the planning phase would be chartered to support the work of the project manager—a role that is quite distinct from that relevant to their work in execution.

Because of its importance, the project plan may be submitted for review by selected key stakeholders (such as functional and divisional managers) before it is formally approved. It is important to note, however, that the approval decision belongs to the owner on behalf of the funder and that the involvement of others is encouraged purely for their engagement.

6.6 Reference Groups and Advisers

As is the situation with the steering committee, the need for reference groups and advisers during planning is peculiar to each project. The roles of any reference groups or advisers appointed during planning is quite distinct from those recognised during execution. During planning, reference groups and advisers are commissioned to work on selected aspects of the plan and they automatically stand down when that work is complete. Typical briefs for such entities cover: risk management, issues management, estimation of critical parameters and compliance with policies, regulations and guidelines.

As is the case with the steering committee, those reference groups and advisers identified to play a role early in execution will be assembled during set up. (Others will be formed as the emerging needs of execution dictate).

6.7 Project Counsellors

If a project assurance counsellor has been appointed before planning starts (or even during planning), he/she would be expected to:

- Support the planning process with appropriate professional guidance to all key players.
- Ensure that the project plan meets relevant quality standards.

If already appointed, the project assurance counsellor would be asked to review the plan and (after discussion with the project manager) table a report for the project owner. Such a report could be based on a simple checklist of criteria, or it could involve more formal methods of assessment. For example, the quality of a project plan could be analysed and compared to other projects using the tool demonstrated in the next section.

If a probity counsellor is to be involved he/she will usually be appointed early execution, but for projects involving significant external expenditures that engagement may start during planning. In that case he/she would be expected to:

- Support the planning process with appropriate professional guidance on procurement.
- Ensure that the project's procurement plan meets relevant commercial standards.

6.7.1 Evaluating the Quality of the Project Plan

Executing a project according to its plan does not necessarily ensure success. If the plan is flawed, the project is unlikely to generate its desired outcomes. While a high quality plan does not guarantee success, it increases the chances that the project will be properly executed and completed successfully. Conceptually, a high quality plan is one that allows a reliable decision to be made about proceeding with the project. The reliability of decisions about a project is reflected in the eventual results of the exercise. Ideally, based on an organisation's project plans, all approved projects would be successful (in terms of management, ownership and investment performance) and all failures rejected although, of course, the latter class will never be known.

In the absence of any knowledge about a project's eventual performance, how is the quality of a plan for a prospective project to be judged? Because this is a quality issue it involves tests. In Chap. 3, we introduced a formal structure for a test, made up of a number of elements, including a set of performance variables. The key issue here concerns the most appropriate characteristics of a plan that enable judgements to be made about its quality. The following discussion contrasts two approaches by Dvir and Lechler on one hand and Zwikael and Globerson on the other.

Dvir and Lechler (2004) assess the quality of the project plan using six criteria related to time and cost. As shown here, these are expressed using the terminology and concepts introduced into Chaps. 4, 5 and 6:

1. That the work breakdown structure is based on work packages at its lowest level (corresponding to "tasks" in the stylised three-level model discussed in Chap. 4). That is, the work breakdown structure takes the form of individually-assignable blocks of work).

2. That every work package has a target date for completion.
3. That an estimate of the slack attached to every component of the work breakdown structure is available. ("slack" being a project management term for "a buffer" of time which can be utilised without extending project duration).
4. That all dependencies amongst all work packages are known and specifically identified.
5. That reliable estimates of outlays for the project are available.
6. That a resource plan is available showing the demand for all key personnel (who does what and when).

Zwikael and Globerson (2004) developed an assessment model, called Project Management Planning Quality (PMPQ), to measure the quality of a project plan. The model is builds on knowledge from the fields of project management, control theory, organisational maturity and organisational support. Based on 33 processes that capture the planning processes conducted by project managers and top management support processes lead by executives (described in Appendix C), the technique was used to evaluate the quality of project plans in different organisations. It can also be employed to identify faulty processes that should be improved in order to enhance planning.

6.7.2 The Quality of the Project Plan in Practice

A benchmarking exercise using the PMPQ approach was also conducted on data from 776 projects to compare the quality of project planning across industries. The average results of each industry on the scale of 1 (low) to 5 (high) are presented in Fig. 6.9.

The above findings concerning the difference across industries were compared with two other studies, which dealt with similar topics. Ibbs and Kwak (2000) evaluated the maturity of 38 US organisations, while Mullaly (1998) evaluated 65 Canadian organisations. Both studies grouped the organisations into industries as

Fig. 6.9 The quality of project planning across industries

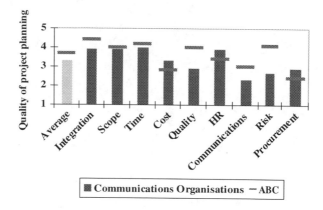

Fig. 6.10 The quality of project planning for "ABC": a communications company

well. These two studies examined all project life phases, while the PMPQ model focused on the planning phase. In all three studies, the ranking of industries across these studies is similar: projects undertaken by construction and engineering organisations produce the highest quality plans, while those assembled by production organisations have the lowest quality.

The PMPQ model supports project planning improvement by benchmarking a single project/division/organisation against others. Moreover, it can identify major gaps in different project management areas. For example, Fig. 6.10 presents a comparison of project planning capabilities in company "ABC" with other companies from the same industry, on average and according to the nine project knowledge areas included in the PMBOK (PMI 2008). The analysis identifies that while "ABC" has higher quality plans than its competitors, there are also areas for improvement such as costing and procurement. Furthermore, the PMPQ approach can identify specific faulty planning processes and trigger improvements in planning practice for future projects.

6.8   Summary

In this chapter, we have considered the structure of a project plan and the processes for producing it. When the project plan is approved, a start can be made on execution. Because the work of producing outputs completely dominates the project environment, the associated management processes can suffer from a lack of attention, effort and interest from key stakeholders. Chapter 7 explores the nature of that critical above-the-line activity.

Chapter 7
Executing a Project: The Roles of the Key Players

7.1 An Outline of Project Execution Management

During execution, team members spend the bulk of their time on below-the-line activities identified in the project plan, while the rest is devoted to above-the-line work that is required to manage the project. We call this latter work, *project execution management*.

Even though they are often involved directly in the production of outputs, project managers are mainly concerned with the above-the-line processes required during this phase. A variety of tools and techniques employed in the project plan provide the "script" that guides execution. For a number of reasons (including the emergence of unforeseen issues and threats), the actual execution will stray from this script, and so a management control framework has to be assembled to monitor, detect deviations from, and allow appropriate revisions to, the script. Such a control framework itself demands a significant amount of activity over and above that devoted to the project's outputs. While the work of producing outputs is unique to the project being undertaken, the associated management and control follows certain common guidelines, conventions and standards (regardless of the project). In what follows, when we speak about "project execution" we are, for the most part, concerned only with above-the-line activities. Some of this work is peculiar to execution (such as the preparation and review of regular project status reports), but some is a continuation of activity that appeared for the first time in earlier phases. A case in point is the identification of threats, one of the processes undertaken as part of risk management during initiation. Although we (somewhat loosely) classify what happens before execution as "planning" and what happens during execution as "management", many planning-type activities continue throughout this next phase.

The forums for execution management are derived from the project governance model specified in the project plan. Two of these are of particular importance: regular management meetings of the project team(s) and meetings of the steering committee. Both forums have standing agendas and (corresponding) standardised

O. Zwikael and J. Smyrk, *Project Management for the Creation of Organisational Value*, DOI: 10.1007/978-1-84996-516-3_7,
© Springer-Verlag London Limited 2011

reporting packages covering the three layers of project execution management: project environment management, project execution control, and project baseline revision (as outlined in the following discussion). In each case, using the relevant reporting package, forum participants decide on appropriate interventions in the conduct of the exercise. These forums and their agendas are covered in the sections below devoted to the project team and steering committee respectively.

On a matter of terminology, although the expression "deliver a project" is common, according to the framework presented, it is not strictly correct, because a project is a process (which represents work). One might *execute a process* (or *deliver its outputs*), but one cannot, in any meaningful way, *"deliver (the work of) a project"*.

7.1.1 Execution Management Processes

Projects are intimately bound up with change so, in a sense, project management is change management. The term "change management" is used rather loosely by the profession to apply to three related, but distinct, concepts: the management of the changes that certain stakeholders face because of the project, the management of the changes imposed by the environment on the conduct of the project and the management of suggested changes to the scope of the project. The first of these is addressed through stakeholder engagement (Chap. 4). The third is discussed later in this chapter. Here we consider the second of these, related to the changes that arise from differences between the way the project was perceived in the plan and the way it develops during execution. Project execution management addresses this particular aspect and involves three kinds of processes.

7.1.1.1 Processes to Deal with Changes in the Project Environment

The environment within which execution takes place is continuously reshaped: internally as the plan evolves and externally through factors such as issues and risks. Project environment management involves processes to deal with emerging circumstances. These processes require continuous monitoring and reporting during execution.

Project environment management relies heavily on four of the elements that make up the anatomy of a project: the three registers (for stakeholders, risk, and issues respectively) together with governance. Each of these is made the subject of a management cycle based on a control loop of four steps:

1. *Monitoring*. Where team members and other key players scan the project environment for significant developments.
2. *Assessment*. Where each development is considered.

3. *Judgement.* Where a decision is taken on an appropriate strategy to deal with the development.
4. *Intervention.* Whereby any required action is planned and implemented.

7.1.1.2 Processes to Control Deviations from Plan

Changes in the project environment can cause certain parameters (such as timeframes and budgets) to deviate from the targets (or thresholds) established in the project plan. Processes are therefore required that attempt to control such deviations. We call this work *project execution control*, which requires that key players have access to variables that can be manipulated as management "levers". Project execution control is effected through three of the elements that make up the anatomy of a project: schedules (timeframes), resources (outlays and/or labour) and scope (outputs/quality). Project execution control is driven by time/cost schedules that are assembled during planning and continuously revised during execution. It too is based on a control loop of four steps:

1. *Measurement.* Where timeframe and cost are measured (for the project to date) and predicted (through to project end).
2. *Assessment.* Where the gap between this measurement and some desired value is calculated (for example, the projected total expenditure at project end versus the approved project budget).
3. *Judgement.* Where a decision is taken on whether or not the gap requires attention (for example, by reducing overtime hours).
4. *Intervention.* Whereby any action is taken to close the gap.

Project execution control also demands regular routine monitoring and reporting so that the longest possible lead time is available for any intervention that becomes necessary to deal with deviations from plan.

Consider a project to extend a freeway between the Central Business District of a city and its outer suburbs that is held up because of bad weather. It is necessary to predict the actual completion date so that the gap between this date and the planned completion date can be assessed. Such a prediction will be based on the duration of work completed to date plus a forecast of the duration of all remaining work. A judgement will then be made about the acceptability of any gap, leading to some sort of intervention (if required), such as using additional premium rate resources.

7.1.1.3 Processes to Revise Baseline Documents

Despite attempts to manage the project's environment and control its execution, circumstances can arise when it becomes clear that a parameter for which targets or thresholds had been set cannot be achieved and so must be changed (in either the business case or the project plan). We call this work, *project baseline revision.*

Project baseline revision is driven by a reporting/review/decision cycle that is maintained throughout execution. It is based on three steps:

1. *Acknowledgement.* When the need to revise the project baseline is accepted and a brief issued to make the necessary changes.
2. *Revision.* When the relevant baseline documents are updated and all of the changes reconciled.
3. *Review.* When revised baseline documents are presented for approval.

Revisions to project baselines are made only as they become necessary and so this component of execution management can be made the subject of exception based reporting. Depending on their significance, changes to the business case may require ratification by the project owner, the steering committee or even the funder.

7.1.2 Accommodating Projects within an Organisational Structure

To provide some sort of administrative stability, organisations commonly adopt "standing" structures (represented by organisation charts). These standing structures define the reporting lines that are used for a variety of purposes, such as identifying accountabilities, setting responsibilities, declaring authorities, evaluating performance and succession planning. Amongst other things, an organisational structure specifies the way in which labour is divided and coordinated amongst distinct processes (Mintzberg, 1979). When processes are stable and repeated (as is often the case with business operations), organisational structures also tend to be stable. In a project environment (where processes are not repeated), then organisational structures tend to be transient. Because they require temporary extensions to the standing structures of organisations, projects present particularly awkward challenges for management.

There are many models for an organisation's standing structures. Roles and people may be arranged using any of a large range of characteristics including: function, product, process, market or geographic location. There is an equally large range of criteria for choosing from amongst the alternative models, including: cost, adaptability, staff attrition and cultural fit. Accordingly, the effectiveness of any particular model for an organisation's standing structure tends to be peculiar to the circumstances of the time—there is no universally "best" model. This leads to a dilemma for management, because considerations of organisational stability require that, once implemented, structural models should not be made the subject of frequent change. The same sorts of issues are raised with each new project but, because projects are temporary, management has much greater freedom in designing an organisational model for each project than it does for its standing structures.

The peculiarities of project structures arise because they are temporary organisations (Müller and Turner 2007), that is, they are disbanded when the project is complete. As a result, project governance arrangements introduce additional relationships and reporting lines that go beyond the organisation's standing structures. Furthermore, they tend to disrupt operational processes by competing for resources. It is for this reason that the business case has a section entitled "Organisational impact".

Two questions must be answered if a project is to be successfully accommodated within an organisation's standing structures:

1. How is the project to be linked into the organisation?
2. How is competition for resources (between projects and operations) to be resolved?

The first question involves the assembly of appropriate linkages between the (transient) structure adopted for the project and the (standing) structure adopted for administrative purposes. Such linkages represent (and allow for) flows of:

- *Information*. From the project to the standing structure.
- *Instruction*. From the standing structure to the project.
- *Authority*. From the standing structure to the project.

By way of contrast, the linkages one finds within an organisation's standing structure also indicate other forms of relationship between nodes, such as performance assessment, reward and career progression.

While it is common to see charts that attempt to locate projects in their organisational context, the value of such diagrams is far from clear. Because of the linkages involved, such figures present some very awkward topological problems for which there are no simple solutions. The creation of (usually highly artificial) arrangements under which steering committees report to committees, for example, may make the diagramming easier, but their contribution to better management is questionable.

How then does an organisation "connect to" its projects? Formal linkages between a project and an organisation are created through members of the steering committee in general and the project owner in particular, as suggested by Fig. 7.1. (This diagram is provided here simply to show where the connections are made between a project and the organisation. It is not recommended as the template for an extension to an organisation chart). Information presented to the steering committee subsequently becomes accessible to the sponsoring organisation and funders. In turn, instructions from the sponsoring organisation and funders are relayed to the project via the steering committee, as are authorities (especially those concerning deployment of resources). Linkages of this kind can connect a project to any number of participating organisations (although only one is shown here).

In addition to these formal linkages, there will normally be numerous informal paths along which information flows out of the project into participating organisations. (Each representative appointed to specific roles will obviously provide such a channel of communication). In most situations it is not necessary (or even useful) to show the linkages between projects and their participating organisations in diagrams.

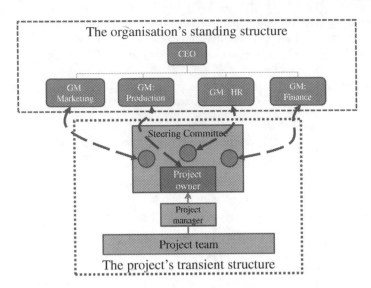

Fig. 7.1 An example of how the project governance model "hooks into" the organisation's standing structure through those who are commissioned to fill project leadership roles

The question concerning resource contention requires some background discussion. The forms of assignment under which staff are appointed to projects can be usefully classified according to the "intensity" of the engagement:

- *Full-time*. Appointments from amongst the staff of a participating organisation whereby the employee disengages from the organisation's standing structure for the duration. (Secondment, full-time appointment for a relatively short period of time, is a special case of this category).
- *Part-time*. Appointments from amongst the staff of a participating organisation whereby the employee remains engaged with the organisation's standing structure.
- *Contracted staff*. Appointments from outside the participating organisations can be either full-time or part-time.

There is a formidable body of opinion and evidence that cross-functional teams are more effective than those having a more narrow skill base. Areas of enhanced performance include external communication, technical quality (e.g. Keller, 2001), creativity (Jassawalla & Sashittal, 1999) and group performance (Pelled, Eisenhardt, & Xin, 1999). Such research can be expected to encourage the engagement of part-time team members, bringing into sharper focus some underlying organisational challenges.

Part-time appointments automatically involve dual lines of reporting in which the staff member is subject to the simultaneous demands of a line manager and a project manager. Even in the case of full-time assignments, conflicting demands can arise, especially if the engagement is for a relatively short period of time. This is how a matrix structure emerges. Note that a matrix structure is not a model of management, but simply a phenomenon arising from multiple lines of reporting. Matrix structures become problematic when they give rise to resource contention. This problem is exacerbated by fluctuating loads and the associated swings in demands for labour from both the project and the standing structure of the organisation. If we use a week as a representative resource-planning horizon, then clearly the total weekly workload on a part-timer from both the project and the organisation will rarely equal his/her availability over that week. When the load exceeds availability, then the employee faces potentially conflicting and irreconcilable demands.

No matter how the team is organised, dual reporting lines bring with them resource contention problems (as reflected in Fig. 7.2)—and so the focus is now on how to manage the resource issues that arise from a matrix structure. While the diagram uses the case of a part-timer, it is completely general in its interpretation—a full-timer can also be faced with a similar problem.

Amongst the strategies to deal with this problem, three are noteworthy. The first concerns the formal acknowledgement of roles identified in the project governance model. All internal appointments should be formalised in a memorandum of understanding (MoU) which serves the same purpose as a contract, but without legal and commercial sanctions. An MoU should establish very clearly key elements of the appointment including:

- The period of the engagement
- The intensity of the appointment (full-time, part-time)
- A nominal level of availability for the project
- Guidelines on how the MOU should be renegotiated
- A strategy for re-deploying the staff member at the end of the assignment

The document would also note the emergence of a matrix structure and acknowledge the possibility of resource contention, as well as outlining mechanisms for dealing with such situations. The MoU would be signed off by the staff member, the line manager and the project manager.

The second approach concerns the relationship between the project manager and the line manager. Resolution of resource contention problems will involve

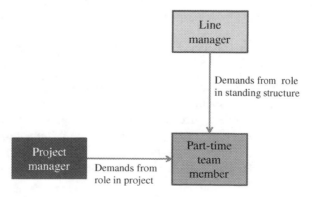

Fig. 7.2 A matrix structure emerging from part-time project appointment— causing resource contention

both players and so the soundness of their relationship will be a significant factor in the effectiveness of their problem solving actions.

The third approach involves a critical role for the steering committee. Participating in the resolution of resource contention problems is a key function of the steering committee and should be reflected in its charter and structure.

7.1.3 Top Management Support

During project execution, project managers and project teams are expected to undertake, manage and control all activities in accordance with the project plan. In situations where these players are drawn from the performing or funding organisations, executive support from their "home" organisations is critical at this time, as it is for overall project success (Fortune & White 2006). As a result, senior line managers should be particularly conscious of the need to support those members of their staff who have been given project responsibility. While in what follows we highlight the importance of providing the project manager with a supportive organisational context, the same comments apply to team members (and also to some reference group members). Consider a project that requires specialist internal staff from a functional unit, but where the functional manager perceives such people as "too valuable to give up". Resolving such an issue in favour of the project requires strong support from senior management.

A supportive and reinforcing organisational context will provide the project manager with the necessary authority and autonomy to make decisions about the project, including those related to recruitment and the management all of the project resources as he/she sees fit. A supportive organisational context would be expected to recognise reward and recognition schemes, training programmes and supporting technologies.

If the organisation is perceived as being ambiguous in its support for the project manager, he or she may be seen as powerless. Managers who are perceived as powerless are severely handicapped in filling their role and, some would argue, doomed to failure. Therefore, gaining and maintaining executive support should be a major part of the project manager's strategy. A recent investigation into top management support processes reveals some interesting results about their adoption across industry groups and countries (Zwikael & Globerson, 2006; Zwikael, 2008a, b). Of seventeen top management support processes analysed (see Appendix D), the most important (in terms of impact on project success) were found to be:

1. *Assignment of an appropriate project manager.* Careful consideration by senior managers should be paid to the match between project and project managers. For example, a successful project manager with experience and interest limited to long-term strategic projects, may be unsuitable for a fast-pace, high-stress project. Factors such as knowledge, experience, project managers' personal characteristics and type of project should be considered when assigning project managers to projects.

2. *Communication between the project manager and the organisation.* Senior managers should display an appropriate level of interest in all projects undertaken by their organisations. Because strategic projects receive such a large share of executive attention, smaller initiatives can be easily forgotten, exposing them to unnecessary risks of poor performance. All project managers must be given the opportunity of regular interaction with the senior ranks of the performing organisation so that their endeavours are recognised and supported. While much of this interaction will be effected through project governance arrangements, it is also important that informal communications be encouraged as well.

3. *Interactive cross departmental project planning groups.* Projects that cross departmental boundaries require special consideration. This may mean, for example, that relevant line managers (or representatives) are expected to participate in project meetings and confirm their ongoing commitment to the project plan.

4. *Project management success measurement system.* Project success measures should be clearly stated by senior managers and acknowledged by project managers. This should be documented in the business case, with explicit targets for time, cost and clear quality criteria for outputs.

7.2 The Project Manager

The project manager is a prominent player in execution because it is here that he/she discharges accountability to the project owner for delivering outputs (fit for purpose, on time, within budget and without causing detrimental outcomes). The accountabilities of a project manager should be accompanied by a suite of responsibilities (for example a responsibility for completing all the work in the WBS) and supported with appropriate authorities (such as the authority to deploy the resources approved to the project). In small projects, the project manager will spend a significant amount of time working directly on the production of outputs. As projects get larger, so does the administrative load. Because the bulk of this falls to the project manager, it follows that the larger the project, the more time he/she spends filling a coordinating/leading/guiding role. Eventually there is a threshold of project size, above which the project manager is unable to shoulder the entire above-the-line load. At that point, it becomes necessary to engage others in this work. One strategy involves the delegation of selected above-the-line tasks to team members, for example by appointment of a risk manager, issues manager and stakeholder manager. Another is to engage a project administrator or, in the case of very large projects, a project administration team.

Care must be exercised in these approaches to ensure that overall responsibility for this work remains with the project manager. The following sections discuss the role of a project manager in these areas.

7.2.1 The Project Manager as a Project Execution Manager

The project manager has considerable responsibility during execution. He/she must be comfortable operating *proactively* in an environment that continuously demands *reaction*. Because the project and its environment evolve relentlessly (regardless of whether the project manager is there), part-time assignments become problematic very quickly. Running projects "off the side of the desk" (with its attendant lack of attention to detail) represents a significant threat to all but the most trivial of initiatives, which should be recognised in every risk register. If project administration is continually competing with below-the-line work for resources, time and attention, then execution management will suffer, with potentially catastrophic consequences.

The project manager must ensure that adequate resources are devoted to each of the three processes of execution management discussed above:

1. *Dealing with changes in the project environment.* By ensuring that the three registers (stakeholders, risk and issues) are maintained, that the associated management programs are being executed, that relevant stakeholders have appropriate access to them and that reliable reports flow to key players.
2. *Controlling deviations from plan.* By ensuring that the foundation instruments of planning and execution (in particular: statement of scope, WBS, Gantt chart, schedules of milestones and budget summaries) are continuously updated, that they are reported accurately to key players, that problems are identified (and advised) early and that recommendations for intervention are formulated quickly.
3. *Revising baseline documents.* By keeping the project owner fully informed of emerging pressure to revise the business case/project plan and providing reliable information on which any necessary changes to project strategy can be based.

Unlike below-the-line activity, if it is not consciously monitored, administration tends to simply "evaporate". Because of this, the project manager has to work hard to maintain the momentum of project execution management. It is here that project administrators can prove very effective. They are concerned exclusively with administrative activity and so it tends to get done in a timely fashion. A dogged, persistent and professional administrator will create a (very desirable) sense of discipline around a project.

7.2.2 Communications Management

Communication was identified earlier as a critical success process. To understand the context within which a project's communications take place, it is necessary to return briefly to the earlier discussion of stakeholder engagement (Chap. 4). One of the three generic forms of stakeholder engagement is "Include in communications plan". This linkage with stakeholder analysis is central to communications management. The project communications plan (and associated communications management) is completely bounded by stakeholder engagement. Looked at another way, the only communication that will be effected in a project is that specified in the communications plan and which, in turn, is defined by the stakeholder register. This linkage between stakeholding and communications is so important that two additional points should be emphasised:

- Non-stakeholders in the project should not be recognised in the communications plan.
- All those stakeholders who are to be engaged will normally be recognised in the communications plan.

The engagement of a project stakeholder invariably implies inclusion in the communications plan. It is difficult to conceive of a situation where a stakeholder is to be engaged in the project but not included somewhere in the communications plan. (It is, however, quite conceivable that although someone has been identified as a stakeholder in a project, subsequent analysis suggests that engagement is not necessary). The project communications plan seeks to address (in a cohesive and systematic way) all the decisions to formally communicate with particular stakeholders.

Throughout execution, considerable time and attention is devoted not only to the ongoing management of communications, but also to the maintenance of the communications plan. This work is underpinned by the continued maintenance of the stakeholder register.

Overall project communication is determined by individual plans which establish the key parameters that will be used to guide the way communications are undertaken for each identifiable stakeholder group. Discussion of these parameters follows with an example from ICO's Project BuyRite where it has been decided that a number of staff in procurement-related areas must be redeployed at the conclusion of the exercise.

- *Key issues* What issues should the communications plan address for this stakeholder group? For example, "Staff are concerned about job security".
- *Why?* What are the desired outcomes from communications with this stakeholder? The two-way nature of communication should be reflected in the statement of these outcomes. For example, "Increased understanding (for ICO) about the depth of staff concern" and "Increased confidence amongst staff that redeployment will leave them better-off".
- *How?* What form will the communication take? This should address choice of media, format and approaches. For example, "One-on-one meetings with an assigned HR case manager", "Access to the Project BuyRite website", "Membership of the (newly created) ICO redeployment club".
- *What?* What key messages are to be conveyed and what critical information do we seek? For example "Your future is assured" and "What is the major cause of your uncertainty?".

- *When*? When will the communication start? How often will it be undertaken? When will it conclude? For example: "One-on-one meetings to start by the end of June, to be available on demand and to conclude three months after redeployment".

In addition to this stakeholder-specific detail, the project communications plan will provide some high-level overall information including:

- Discussion about the general issues that dominate the communications landscape, to provide an overview of the context within which the plan is being formulated. This discussion is centred on general background issues of common interest to all stakeholders in the project.
- An overarching communications strategy is expressed as a cohesive portfolio of communications mechanisms. If, for example, six separate needs for newsletters are revealed in the analysis of different stakeholders, it would be highly desirable to address these through a single publication.
- A workplan covering both the development and delivery activity. For the newsletter this would map out the work involved in:

  - Developing the publication.
  - The regular cycle of writing material, publishing and distribution.

- Estimates of resources required for development and delivery.

During execution not only is all of the development and delivery activity undertaken, but (as new stakeholder information becomes available), the communications plan is continuously updated.

7.2.3 Risk Control

Work to mitigate the project's risks was covered in some detail in Chap. 6. The processes described there continue throughout execution, as one of the mechanisms to support management of the project environment. It is useful to summarise what would have been happening with risk in the project to date.

- During initiation, selected risks would have been taken through to the point where a mitigation strategy had been agreed.
- During planning, newly emerging risks are also processed in this way. In addition, detailed plans are assembled for the work required to implement all agreed mitigation strategies.

All this implies that, during the execution phase of the project, not only are all mitigation plans implemented, but also the risk management process (in its entirety) continues as an ongoing activity. A team member will have been appointed as risk manager (a role that, by default, falls to the project manager) who will manage all risk-related activity.

Three sorts of event trigger the risk management process during project execution, and each involves a peculiar variant of the risk management process:

1. The spontaneous emergence of a new threat. For example, we now believe that our only two suppliers of concrete additives may merge, leading to a monopoly.
2. A scheduled risk review session when, amongst other things, teams or individuals proactively search for new threats.
3. The realisation of a triggering event (regardless of whether or not it had been identified in the risk register at that time). For example, the consultant who has been doing the business process analysis accepts an offer to join one of ICO's competitors.

7.2.4 Issue Management

Issue management is another of the processes to deal with changes in the project environment. An issue is defined as a matter of general concern to the project (other than a risk) that requires a response, action or resolution. Many issues take the form of either a task that would otherwise have been overlooked, or a question that needs to be answered. Consider a well-advanced project involving a team in London and a subcontractor located in Milan. A new airline announces the introduction of an additional service between the two cities. That is an issue for the project because there may be implications for the workplan (and perhaps even the project's cost). Issues can range in importance from critical to trivial. Part of the skill of issue management is knowing how to sift out those that can be safely ignored.

If issues are not proactively managed, they can have either of two undesirable consequences for the attractiveness of a project. In the first case they can evolve into a risk (thus exposing the project to potentially damaging impacts). In the second they emerge as opportunities which will be lost if not addressed. Issues are included in the regular reports by the project manager because key players should be aware of those issues to which the project is particularly sensitive.

We have adopted this particular definition of "issue" because it gives rise to a simple, predictable useful technique of issue management. Under this approach, issues can evolve into risks, but not vice versa. While there is an *analytical model* for risk (based on the event-impact mechanism), the management of issues is a simple *clerical procedure*, supported with an issue register (described below). Although we use the concepts of risk and issues in distinct ways, an issue register is somewhat akin to a risk register in structure, and fills a similar role. A team member will have been appointed as issues manager (by default the role falls to the project manager) who will manage all issues-related activity.

7.2.4.1 Issue Management Processes

Issue management involves five processes, as outlined in Table 7.1.

Table 7.1 The issue management processes

Issue management process	Project phase	Outputs	Leader
Identification	Initiation (ongoing)	A list of issues	Champion
Analysis	Initiation (ongoing)	Issue register	Champion/Project manager
Management planning	Planning (ongoing)	Issue management plan	Project manager
Management implementation	Execution (ongoing)	Managed issues	Project manager
Issues monitoring	Execution (ongoing)	Updated issue management plan	Project manager
		Updated issue register	Project manager

7.3 Box 7.1 Practical Considerations

The Phenomenon of the "Exploding Issue Register"

The size of issue registers (gauged by the number of active entries) varies widely between projects. Size is influenced by a number of factors, including:

- The diligence (or zealotry) of the issues manager
- The novelty of the project
- The completeness of the WBS

This last item bears further discussion. If the WBS is incomplete or unreliable, then tasks that should have been identified there remain hidden until they can no longer be avoided and require immediate attention. The "sudden" emergence of such tasks makes them appear as issues and so the issue register rapidly expands with items that are really missing tasks from the WBS.

If an issue register begins to grow rapidly, the project manager should investigate whether or not this has been caused by an incomplete or unreliable WBS. If this is found to be the case, then clearly the project's workplan will need thorough revision.

Issues that have been identified during project initiation appear in the first version of the issue register. That version of the issue register is then incorporated into the business case. After an appropriate update during the planning phase, it then appears in the project plan.

Issue management begins as soon as the very first work starts on a business case and continues throughout the life of the project. As is suggested in the earlier coverage of the initiation spiral (see Chap. 5) an issue register is often the very first formal piece of project documentation, assembled even before the first cut of the scoping statement, to serve as a repository for items that have yet to be given a more permanent home elsewhere in the project documentation.

Two sorts of event trigger the issue management process during project execution, and each involves a peculiar variant of the risk management process:

- The emergence of a new issue.
- A scheduled issues review session when, amongst other things, the team proactively searches for new issues.

The issue register is a formally structured catalogue of the issues that arise during the project, also covering details about the actions that have been adopted to resolve each. Numerous formats for issue registers are offered in various project management toolkits and methodologies.

7.2.5 Schedule Control

The project manager is responsible for project execution management, especially the processes to control deviations from plan. The first two of the four steps that surround these processes are measurement and assessment, both of which require supporting tools. A complication, arising from the interplay of timeframe and expenditure, makes some commonly used naive approaches all but useless in this role.

Consider, for example a simple graph on which planned and actual expenditures are plotted cumulatively against time like that shown in Fig. 7.3. Note that, because it is based on a known schedule (covering the entire duration of the project's execution), the "Planned" graph is complete. The maximum (at the right-hand end) is, of course, the project budget. The "Actual" graph, on the other hand, is known only up to the date of reporting (the end of June in this case) and so it is "incomplete"; the remaining months have, as yet, no actual values.

What does this graph tell us about the health of the project? A simplistic interpretation seems to suggest that the project is (desirably) below budget, but a little thought quickly reveals that it tells us absolutely nothing about the financial

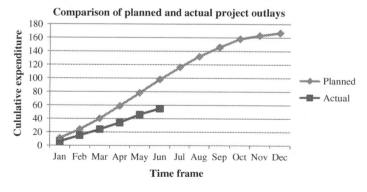

Fig. 7.3 Naïve project expenditure tracking graph

state-of-health of the initiative. Two totally conflicting scenarios are completely consistent with the graph that we see:

1. *Poor progress, excessive expenditure*. The project team is exceeding the budget on every cost it has incurred to date. At the same time it has made very little progress against its plan (and accordingly has not made all of outlays that were originally expected by this date).
2. *Rapid progress, low expenditure*. The project team is not finding it necessary to spend as much as had been budgeted for all of the outlays it has made to date and, at the same time it has made such rapid progress that it is well ahead of plan.

7.2.5.1 The Earned Value Methodology

Because the graph cannot discriminate between these contradictory situations, it must not be used (or at least not in the form displayed here). A number of techniques can be employed to overcome this (fatal) flaw, for example:

1. Base the graph on milestones (rather than dates).
2. "Complete" the actual graph by projecting it through to the end of execution. This requires that values be provided for the future months by replacing (unknown) actual values with (estimated) forecasts.
3. Analyse scope together with time and cost.

This latter approach is the basis of the Earned Value Method (EVM). Earned value is simply a way of analysing actual cost, planned work and actual work for tasks in the WBS in order to gauge the project's state-of-health as far as timeframe and budget are concerned. It is tempting to read something of significance into the name, which suggests that it introduces a new concept of "value". EVM simply provides information about dates and outlays and provides predictions of their final values at the end of a project. Although the technique can be applied to outlays and labour separately, it is most effective in financial tracking.

Recently, following shortcomings of EVM in forecasting the actual project duration, Earned Schedule (Lipke, 2009) yields time-based indicators. It analyses the gap between the actual and the planned times attached to scheduled levels of expenditure.

7.2.6 The Project Manager's Role as a Team Leader

As well as the management of all activity (regardless of whether it is administration or production in nature), the project manager plays a critical role as leader. Human Resource Management (HRM) related activities have significant impact on project success. The research study described in Appendix C provides evidence in support of this view. It was found that of all executive planning processes, the appointment of an appropriate project manager contributes most significantly to project success, while staff acquisition is the most important project planning process conducted by the project manager. Chap. 6 discusses in more detail some of the HRM activities related to the planning phase of projects.

In light of this, we review the relevant HRM activities in which project managers are engaged during execution and explore some of the challenges that they face. We then go on to suggest a number of supporting tools and practices.

7.2.6.1 Leading Teams

In the transition from planning to execution, project managers find that their leadership role becomes increasingly important. As well as the growth in team size, other factors are often at work here, related to differences amongst team members in background, interest, motivation and objectives. In filling this role, the project manager may seek functional support from the organisation in areas such as recruitment, induction, formation and redeployment. However, it is the project manager's responsibility to assemble and motivate a workable and effective team. To do that, special attention must be paid to the following issues:

1. *Team member assignment.* The project plan provides important information about the demand for resources on which the make-up of the team will be based. When appointing team members, a number of criteria come into play. In addition to technical expertise (which is essential), candidates must be suited to working in groups. This is particularly important when the role entails high levels of interpersonal communication, coordination and high levels of task interdependency.
2. *Team size.* The project plan will provide clear details of the skills required during execution, as well as estimates of the quantities of each. For a given level of resourcing, there may be some flexibility in deciding on the size of the corresponding team. For example, there may be options based in the ratio of full-time to part time and seconded staff. Selection of an appropriate option will be constrained by the fact that as a team gets larger (measured by head-count), so too do the problems of communication and coordination. All of this is moderated by the quality of the resource estimates in the project plan. The higher the quality of the resource estimates in the project plan, the easier it will be to decide on appropriate resourcing levels and team size.
3. *Team member mix: homogeneity or heterogeneity.* A final issue to be addressed in the composition of a team relates to the level of similarity or difference among team members. Despite the current general interest in (and vigorous promotion of) diversity, team member homogeneity (as reflected in similarity of thought and behaviour) may support coherence, consistency and achievement of consensus. On the other hand, a lack of diversity may constrain the team's creativity. If not handled in a sophisticated way, heterogeneity can prove

problematic. Given enough diversity amongst team members, convergence of ideas or agreement can prove almost impossible to achieve.

Transforming a group of individuals into a cohesive team is a challenging task, especially in the case of members who are seconded or allocated part time to the project from elsewhere within the organisation. The project manager will have to invest considerable effort in creating conditions that will foster such a transition.

The leadership, management and coaching of a team goes beyond issues of technical competency, it is also dependent on the project manager's interpersonal and people-oriented behaviours (Cannon-Bowers & Salas, 1998). In order to get the most out of technically skilled team members, the project manager must be a skilled and proficient communicator and motivator. In the sections below we discuss the crucial motivational role of the project manager and examine the role that organisational context plays in supporting these efforts.

7.2.6.2 Motivating Team Members

Motivating an individual to be a fully committed member in a project team is not an easy task. Being assigned to a project may be perceived by some as inconvenient at best, or even as career-limiting at worst. This latter point may become a significant issue in an organisation where advancement and promotion is decided by operational performance, especially for full-time project assignments. In this case, individuals may become extremely reluctant to join a project because they become effectively "off-line and out-of-sight". The fear of being seen as "replaceable" and uncertain about job prospects at the end of a project engagement may even cause resentment. If team members do not perceive the project as a challenging opportunity that will enhance their individual, professional, or social standing, it may become impossible to engage them in a meaningful way.

7.2.6.3 Engaging Team Members

Like other stakeholders in the project, team members need to be considered when formulating engagement programs. Involving prospective team members in planning can be an effective strategy here because it provides an opportunity to introduce them rapidly to the project, helps them understand what will be expected of them and enables them to come to grips with the roles that others will play. Encouraging team members to provide input and influence project decisions is a useful first step in developing a sense of responsibility and commitment. Participation in this sort of activity can also help in development of the working relationship between team member and project manager.

Workshops involving the entire team can also be extremely useful in building a shared vision for the project, clarifying roles and setting expectations. Participation of team members in decision-making processes has been found to have a positive effect on their commitment, motivation and performance (Latham et al. 1988; Rodgers and Hunter 1991).

7.2.6.4 Setting Goals

So that the behaviour of the team is aligned with the project, every member should have an agreed set of goals, supported by appropriate measures and targets. These should reflect not only the performance of each individual, but also the team as a whole. Targets are a strong motivating force, as they encourage people to compare their present capacity to perform with that required to succeed. If employees believe they can achieve their goals, they will tend to persevere until they are successful (Locke & Latham, 1990). To be effective, performance measures should (Rodgers & Hunter, 1991):

- Be defined, monitored and evaluated
- Have targets that are difficult and challenging, yet feasible
- Be decided upon and set in a participative manner
- Include a date for achievement
- Provide the basis of feedback

Details about goals, targets, performance measurement and reward should all be included in the brief used to engage each project participant.

7.2.6.5 Disbanding the Project Team

The role of the team ends when it delivers the project's outputs. If a project's outputs are to be delivered progressively over an extended period, then members of the team (or even entire sub-teams) may be stood down progressively. Depending on the nature of the assignment, standing down a team member from the project can take any of a number of forms:

- Appointment to another project.
- Return to a full time operational role (that had been made temporarily part-time by the assignment).
- Return to normal working arrangements (that had been impacted through secondment to this exercise).
- Appointment to a new role in the organisation.
- Release from the organisation (normally agreed at the time of the appointment to the exercise).
- Termination of a contract (when the appointee was drawn from outside the organisation).

The way in which each of these stand-down processes is conducted can influence the individual's motivation to join another project in the future.

7.2.7 *Output Closeout*

One of the key tenets of modern business management is that processes can be and should be improved over time. The whole quality movement can be viewed as an expression of this principle. Operational processes have been the main target of this sort of attention, for which the management profession has developed a variety of strategies, tool and techniques. Continuous improvement is one of the most important of these strategies, whereby a process is periodically and incrementally adjusted so that its performance is enhanced progressively over time.

Operational processes are repeated, a feature that is exploited in the techniques of continuous improvement. The learning from previous executions can be carried forward into later executions. Because projects are non-repeated processes, it is necessary to adapt the concepts of continuous improvement before they can be applied usefully to the enhancement of project performance. While the value of this sort of approach is widely recognised, it is, unfortunately, rarely used. As a result, we see the same mistakes being repeated, not only by the one organisation, but even by the same project participants.

Closeout is a formal process by which the organisation can learn from each project experience and enhance its future performance. The process is conveniently based on a workshop and results in a report for action by the funder (and other key players). In most projects there will be two variants, identified respectively as *outputs* closeout and *outcomes* closeout. *Outputs* closeout is conducted in either of two scenarios: when execution has been completed and when a project is aborted. An outputs closure workshop should be arranged as soon as is feasible either after the decision to close the project or following delivery of the project's outputs. Outcomes closeout is conducted as soon as the flow of target outcomes is secured (which is usually much later). The process for both variants is the same, differing only in the areas of project performance that are selected for review. In Chap. 8 we describe that process in some detail, confining our attention here just to the list of performance areas that are relevant to output closeout.

Closeout is quite distinct from the coverage of a conventional "project review" in being strictly above-the-line. By way of contrast, a traditional "Post Implementation Review" (PIR) is usually concerned with making sure that the outputs are working as intended (which is a below-the-line concern). A PIR should be completed as part of the implementation of an output. According to the framework adopted here, implementation is the third step in creating an output (the other two being production and delivery). Outputs closeout, therefore, *follows* any PIR that might be conducted, and so one is not a substitute (or synonym) for the other because both have very important (but distinct) purposes.

In a large project, where outputs are being delivered by a number of teams over different timeframes, then an output closeout may be conducted as each major deliverable is implemented. There are certain forms of project where closeout may be delayed until sometime after delivery of an output. This is the case in some forms of construction project where a handover of the structure to a client may not take place until a defined period of operational use has elapsed. Outputs closeout is also undertaken if a project is aborted. It is important that the closeout workshop be conducted quickly because as outputs are delivered, members of their teams will stand down making further involvement difficult. Participants in the closeout workshops should be chosen carefully, they must be able to make considered judgements about how well things went (unencumbered by personal impacts and political issues).

The overall approach to closeout borrows heavily from the continuous improvement movement and involves the following steps:

- Select the areas of performance for review.
- Evaluate the gap between desired and actual performance.
- Identify the factors that caused the observed level of performance.
- Extract the learnings from the analysis.
- Decide on the implications for future projects.

Table 7.2 offers a generic list of output closeout performance areas.

On larger projects, it may prove useful to have the closeout workshop facilitated by someone who can exercise a satisfactory level of independence (such as the project assurance counsellor), but on very small exercises, a pragmatic approach would be to have this led by the project manager. As is pointed out in the discussion of closeout in Chap. 8, closeout is not a "post mortem". In projects that have been aborted (or that are seen as failures already), it requires considerable sophistication on the part of the facilitator to extract value from the process. Unfortunately, for many key players, the embarrassment of failure serves as a strong disincentive to conduct closeout workshops, thus making the organisation particularly vulnerable to repeating the same mistakes in the future.

A final comment about performance evaluation is appropriate. The closeout process should not be used to judge individuals (let alone assign "blame"). Judgements about individual performance should be encompassed with the organisation's human resource framework. In particular, at the end of execution, but outside the closure process, the project owner should comment on the performance of the project manager.

7.3 The Project Team

While the project team is primarily concerned with producing project outputs, members will also have demands on them to participate in various above-the-line activities. Candidate processes for delegation by the project manager to the team include issue management, risk management, stakeholder management, meeting administration (agendas/minutes), WBS/Gantt chart maintenance, progress reporting and budgeting. Similar comments apply here to those made when introducing the project manager as execution manager. When team members are delegated administrative roles, they face an internal resource allocation problem

Table 7.2 A generic list of output closure performance areas

Performance area	Focus question for evaluation of the performance area
Approved business case	How clear was the approved project business case and how reliable did it prove to be?
Revised baseline document quality	How clearly did the most recent approved project plan set direction for the work of the project?
Estimation	How reliable were the estimates of: outlays, internal labour and duration?
Internal resource estimates	Did the demands on staff accord with our labour estimates?
Outlay estimates	How reliable did our estimates of cost prove to be?
Risk management	How well did we manage the risks we faced?
Issue management	How well did we manage matters of "general concern" that arose?
Outlays	Did we adhere to the budget for our expenditures?
Delivery Timeframe	Did we deliver our outputs by the agreed dates?
Governance	How well did the governance model work—and in particular the Reference Groups?
Management	How well did our project management arrangements work?
Tracking	How closely were we able to monitor the project and influence developments?
Reporting	How useful were the progress reports tabled during the project?
Scope (ex ante)	Did we deliver all the outputs that were in scope—and did they meet their quality specifications?
Stakeholders	How well did we engage stakeholders?

that requires both leadership and support from the project manager. Even when a project is being serviced by dedicated administrators, there is still a significant above-the-line demand on team members, who often see that sort of activity as a distraction from, or interruption to their "real" job (of producing deliverables).

7.3.1 Formalising Team Roles

One of the seven principles of project governance (Sect. 4.3) is that everyone's role in the project should be formally described and acknowledged in some type of appointment document. Goals, targets, expectations and performance measures, once agreed and accepted should be included in such an agreement, as should be the details of any agreed stand-down arrangement.

7.3.2 Regular Team Meetings

Project teams should meet weekly (in relatively short sessions) to discuss the same topics that are suggested below for the steering committee. Typically, these meetings review the past week, preview the following week and "glance ahead" at the rest of the project. It is neither necessary (nor is it desirable) to have all of the project's routine above-the-line activity scheduled for these sessions. The bulk of regular administrative work will be done offline by individuals or small working parties meeting as and when required. The team meetings should focus on project-wide issues, and topics that require consensus or general discussion. Team members will, of course, be meeting frequently in the normal course of their below-the-line responsibility, but it is desirable that management and production meetings be kept separate.

7.4 The Project Owner

The owner has a number of different roles to fill over the life of the project. During execution, he/she plays the part of *client* to the project manager and so they both become preoccupied with delivery of outputs, the project owner as "purchaser" (who requires these outputs because they are necessary for outcome generation) and the project manager as "provider" (who is then accountable for supplying these outputs). Thus, both are intensely interested in the criteria for project *management* success that were introduced in Chap. 3. Just as the project manager seeks to deliver all outputs fit-for-purpose, on time, within budget and without any detrimental outcomes, the project owner should seek continuous reassurance that all this will happen as planned. Not surprisingly therefore, a close working relationship during execution will make life easier for both.

The project owner should see him/herself as a supporter, trustee and problem solver for the project manager. At the same time, as the project manager's client, the owner, should be continuously concerned (almost to the point of obsession) with the quality of everything delivered by the project manager and team (both above- and below-the-line). When it comes to decision-making, for example, the project owner must judge the quality of reports on their reliability. Unfortunately, in practice project owners often rate the acceptability of a report based on *the level of optimism* it displays rather than on its *plausibility*. The project management profession is only just beginning to realise what the medical profession concluded a long time ago— that decisions must be based on evidence, not on wishful thinking. If a project assurance counsellor has been appointed, then much of the responsibility for judging the quality of reports will fall (appropriately) to that person.

7.4.1 Managing Scope Change

Chapter 1 discussed change management in general. Sect. 7.1 formalised the management of changes that arise during project execution. Here we consider the management of change, as it relates to scope.

Sometimes, circumstances make it necessary to consider changing the project's scope and so two questions arise: "When is a proposed change in scope appropriate?" and "What should be done to the current business case (or project plan) following an agreed change in scope?". A key role of the project owner is to ensure (in consultation with the project manager) that any changes in the scope of the project are handled thoroughly and carefully.

Scope change, as reflected in revisions to lists of outcomes and outputs, will inevitably have important consequences for the project. Current accepted project practice lays considerable emphasis on "scope creep", the tendency for projects to grow over time through the addition of outputs to those agreed at the outset, a phenomenon which is claimed to be a significant cause of project failure (e.g., Cui & Olsson, 2009). At the same time, projects that get into trouble are often subjected to arbitrary and radical "descoping", reductions in scope in order to meet time and cost constraints.

The current scope of a project can be challenged in two ways. In the first, a new output (or an additional fitness-for-purpose-feature) is proposed for inclusion, while in the second an existing output (or an existing fitness-for-purpose feature) is proposed for deletion. Management of the proposed change depends on whether it involves an ITO, or non-ITO output. The first step in the process requires that the proposer of the change identify the additional (or deleted) output according to the taxonomy introduced as Chap. 5. The treatment then varies according to this classification (in which, in addition to ITO-outputs, we recognise four types of "non-ITO output):

1. *An ITO-output.* In this case, the utilisation map provides the foundation of the analysis. To accept an additional ITO output it would have to be shown that either the target outcomes are not currently achievable, or that a new outcome is to be targeted. To delete an ITO output it would have to be shown that either the target outcomes can be achieved without it, or that an existing outcome is no longer to be targeted. If, after thorough analysis of the utilisation map, it is found that a change in scope is appropriate, then the business case must be re-examined, re-worked and reconsidered, to ensure that the revised project still makes sense.

2. *A risk mitigation output.*

   – An additional risk mitigation output may be required if a new threat emerges, or if a current threat requires further mitigation.
   – An existing risk mitigation output may be deleted from scope if a current threat evaporates, or if it no longer merits the creation of a mitigating output.

3. *A stakeholder engagement output.*

   – An additional output may be required to engage a newly-identified stakeholder, or if the engagement plan for a current stakeholder is found to be inadequate.
   – An existing output may be deleted from scope if an entity's stakeholding in the project diminishes.

4. *Mandatory output.* Changes in law, regulation or policy can require additional outputs of this kind, or, equally, may allow an existing output be deleted from scope.
5. *Dependent output.* Changes in the scope of the project can have "knock-on" effects. The addition of an output (regardless of type) to the project's scope, may require that another output be produced. Likewise, the deletion of an existing output from scope may mean that another is no longer required as well.

If the proposed variation in scope cannot be validated using any of these approaches, then it must be classified as arbitrary and rejected out of hand.

One situation that leads to almost irresistible demands for a reduction in scope is worthy of note. Although all of the parameters for a project are clearly stated in the business case and project plan, additional constraints on the agreed timeframe and budget can emerge for either of two reasons:

1. During execution, the dates on which milestones are achieved start to run late, or the expenditures attached to each start to exceed the budget.
2. The funder has arbitrarily reduced the timeframe and/or budget.

In either event, the project becomes infeasible (that is, the agreed scope cannot be delivered within the budget and timeframe). A response is to descope the project by removing outputs (or fitness-for-purpose features of committed outputs). Descoping has the effect of reducing both timeframe and costs, often dramatically. At first glance, this appears to be a very desirable effect (because as costs fall, worth rises), but the harsh reality is that the loss of outputs (and/or fitness-for-purpose features) will cause outcome generation to weaken, inevitably leading to fall in worth. The utilisation map will indicate (even if only qualitatively) the way in which target outcomes will respond to descoping. A priori reasoning suggests that, under normal conditions, removal of outputs will trigger a catastrophic collapse of outcomes relatively quickly, while reduction in the quality of outputs (by removal of fitness-for-purpose features) will cause a progressive fall in outcomes (before they too eventually collapse to zero).

7.4.2 Scope Change and Risk

The impact of scope change on risk must be considered when re-appraising the business case.

Consider, for example, a project with an initial level of worth and risk as is indicated by point "A" in Fig. 7.4. In general, any change to the project will cause corresponding changes (positive or negative) in both worth and risk. The two possible directions of movement for the two parameters (shown as arrows originating at "A") divide the plane into four sectors. These are labelled as "W", "X", "Y" and "Z". We can make some general observations about proposed changes to a project, based on the region into which it then moves:

Fig. 7.4 Trading-off between the worth of a project and its risk

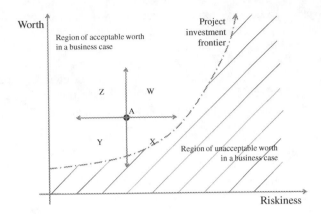

1. *W* requires a trade-off between an increase in the worth of the project and its increased riskiness. Take, for example, an approved business case that assumes a particular risk mitigation programme (and allows for the associated costs). If the funder now decides to abandon some elements of the original risk mitigation programme, then the revised business case would represent a move into region W.

2. *X* represents an undesirable change, because not only does the change to the project force a reduction in its worth, but it also increases risk at the same time. Such a change might, nevertheless be approved in certain circumstances, such as where a budget constraint suggests not only that the quality of ITO outputs be reduced, but that some risk mitigation be abandoned as well.

3. *Y* involves a trade-off between reduced worth and reduced risk (the opposite trade-off to that required in W). This is the region into which a project is pushed by expanded programmes of risk mitigation.

4. *Z* represents a desirable change, requiring no trade-offs. Not only does it increase the worth of the project, but it also causes risk to fall at the same time. Despite the attractiveness of such a change, it may not be feasible in light of budget and/or time constraints.

7.4.3 Managing Schedule Change

Changes in the scope of a project forces changes on the schedule of milestones, reflecting what happens when the plan is assembled for a new project. Other circumstances (unrelated to scope change) will also trigger changes in the schedule of milestones for a project. One of these relates to errors in the original schedule. This arises when early progress on the project makes it clear that the schedule is hopelessly optimistic and that one or other of the timeframe or budget will have to be revised if the project's outputs are to be delivered fit-for-purpose. The project owner must be especially alert to trends that indicate an unreliable project plan, so that he/she can step in quickly to find the causes of the problem and decide on appropriate corrective action. While on one hand the project owner wants to see the plan achieved, he/she does not want to discover that the project is infeasible when it is too late to recover from the situation.

A (surprisingly) little-known device serves as a powerful instrument to gauge the quality of a project plan. The Milestone History Monitor (MHM) described by Philips (Krooshof, Swinkels, & Van der Wal, 1992) is used to monitor changes in the dates of key milestones to warn of "pathological" timing problems. The MHM concentrates on the way in which key milestones have been rescheduled in the course of the project. Slippage of milestones is a common occurrence, while the advancing of milestones is relatively rare. Slippage is not necessarily "pathological"—although it can lead to the complete failure of a project. This sophisticated approach involves tracking *slippage* in milestones and monitoring the resulting patterns of revisions (which are displayed graphically). Regardless of the particular tools involved, the project owner must treat all slippages of milestones as symptoms of fundamental problems, especially when it becomes clear that, through slippage, the original schedule of milestones is infeasible. In that case, the project baseline must be revised and reconfirmed as acceptable. The project owner has every reason to become alarmed if, on reporting that a project has drifted significantly over budget or over time, the project manager declares "But we are going to make it up!". Such an announcement implies that the remainder of the plan is irrelevant and so now can be disregarded.

7.5 The Steering Committee

The steering committee is made up of a small group of key stakeholders, led by the project owner, and chosen because they are both influential and supportive. To fill its primary role of ensuring that the project is executed according to the business case and the project plan, the steering committee must closely monitor progress in regular, formal meetings. These meetings are guided by a relatively simple agenda, at which a standardised report is considered. Some of the items in that report will be for information (especially if the project is progressing satisfactorily), while other items will require decisions and actions. Steering committees typically meet monthly, but this will vary according to the size and importance of the project, as well as the point reached in its life.

7.5.1 A Stylised Reporting Package

In steering committee meetings, the project manager reports on the progress of the project. Clearly, the project manager and project owner must liaise closely

Box 7.2 Practical Considerations

Getting Steering Committees Working

Anecdotal evidence suggests that, in general, steering committees do not meet as often as their role demands. One common reason given for this is that the same senior people are on all project steering committees and so they are heavily constrained in the amount of time they can give each project. Such a situation can reduce the effectiveness of a steering committee through effects such as:

- Irregular (or missed) meetings.
- Loss of continuity (amongst steering committee members).
- Poor levels of understanding amongst steering committee members about the status of the project.
- Inconsistency in the guidance and instructions from the steering committee to the project manager.

A partly related issue concerns the levels of skill amongst steering committee members to fill their role. Because they are often senior, experienced staff who have achieved their positions through a history of successful operational performance, it is assumed that they are automatically qualified to take on their project responsibilities. In practice, it appears that such stakeholders are encumbered by misconceptions about what a steering committee does, lack of awareness of their own roles and poor understanding of how projects should be undertaken.

All this exposes projects to unnecessary risks, which demands a well-considered response. Organisations are encouraged to formulate supportive policies that (amongst other things) might state positions on:

- The importance of steering committees.
- Meeting the executive load from steering committees.
- The obligations of members.
- Judging steering committee performance.
- Competency and capability standards for members.

between meetings. It is important that they both support the thrust of the report and so one would expect that normally, the contents of each regular report would be agreed between them before it is tabled.

The project execution report has three sections, corresponding to the three processes described in the discussion of project execution management provided in Sect. 7.1 , with a structure along the lines of that suggested Box 7.3. It is important that regular reports be kept simple and short, without any superfluous detail. Long reports are undesirable for three reasons:

Box 7.3 Tools for Steering Committee Meetings

A Suggested Template: A Project Execution Reporting Package

1. *Introduction*

   1.1. *Purpose of report.* Simple, stylised statement that appears in every report.

   1.2. *Executive summary.* Summarises major observations, conclusions, issues and items requiring decision or action.

   1.3. *Structure of the report.* Simple overview of how the report is structured—to assist newcomers to navigate the document.

2. *Project environment management*

   This section of the report advises stakeholders about important aspects of the project environment, summarises (and seeks approval where appropriate) for recommended programmes of management.

   2.1. *Stakeholder report*

   - Summary of significant changes since the last report.
   - Extract from stakeholder register—showing all significant changes since the last report.

   2.2. *Issues report.* An extract from the issue register, based on two filters:

   - Issues above a set threshold of importance.
   - Issues assigned for resolution to members of the Steering Committee.

   2.3. *Risk report.* An extract from the issue register, based on three filters:

   - Risks above a set threshold of risk exposure.
   - Risks whose risk exposure has moved by more than a threshold.
   - Risks where mitigation involves members of the Steering Committee.

3. *Project execution control*

   The control of time and budget involves the acknowledgement of three horizons:

   - Project to date, using actuals (historical data).
   - To project end, using forecasts (future data).
   - At project end, using projections (actual + forecasts)

   3.1. *Time frame.* A project progress report, based on the current schedule of milestones, with four short sections.

   - Milestones due for achievement since the last report and confirmation (or otherwise) of their achievement.

- Milestones due for achievement before the next report and confirmation (or otherwise) of their achievability.
- Implications for the project overall of the current rates of progress.
- Discussion of any significant deviation from the currently-approved workplan and proposed strategies to manage those deviations.

3.2. *Budget.* Depending on the way in which the business case and project plan were framed, this section may cover none, one or both of:

- Outlays.
- Labour. Discussion of any significant deviation from the currently-approved budget and proposed strategies to manage those deviations.

3.3. *Deliveries*

- Confirm all deliveries effected since the last report and table quality certificates for each.
- Confirm all deliveries to be effected before the next report.

4. *Project baseline revision*
This section restates the key parameters that are to be recognised from this point on in the project, highlighting any which require change and formally seeking approval to accept the changes. A table format is useful with one row for each parameter and columns for: last approved value, new approved value, rationale for the change. The parameters to be reaffirmed or changed are:

- Target outcomes (with full definitions)
- Outputs (with lists of fitness-for-purpose features)
- Undesirable outcomes (with descriptions)
- Budget (outlays and labour)
- Timeframe (as a schedule of milestones)

- *They require long lead times* not only so that steering committee members can read them, but also so that the team can write them.
- *They are out of date* by the time they are tabled (because of the long lead time).
- *They are expensive* because they involve excessive above-the-line resources.

Box 7.4 Tools for Steering Committee Meetings

A Suggested Agenda

1. *Project environment management*

    1.1. *Stakeholders.*
    1.2. *Issues.*
    1.3. *Risks.*

2. *Project execution control*

    2.1. *Time frame*

       • Review project progress report.
       • Decisions about strategies to manage deviations.

    2.2. *Budget*

       • Review project budget report.
       • Decisions about strategies to manage deviations.

    2.3. *Deliveries*

       • Acknowledgement of recent deliveries.
       • Acknowledgement of the achievements of the team in effecting these deliveries.

3. *Project baseline revision*

    • Review of all baseline parameters.
    • Decisions about changes in baseline parameters.

7.5.2 A Stylised Agenda

Meetings of the Steering Committee are strictly above-the-line, that is, they are confined to discussion of the managerial aspects of a project. This implies that it cannot be assigned (nor should it undertake) any tasks associated with production, delivery or implementation of outputs. This separation of role does, however, become rather blurred from time-to-time. Take the example of a project to redevelop an old integrated iron and steel plant site as a multipurpose port. Tenders have been received to undertake the extensive (and expensive) civil work required to deal with significant ground pollution from the original coke ovens. Analysis of these bids shows a wide variation in prices, timeframes and risks. It is appropriate that when the steering committee makes a decision, it will not only consider costs, timeframes and risks, but also the criteria being used to rank the options and the technical analysis carried out by the project team. The latter topics are, in reality below-the-line, but any attempt at preventing such a discussion would be unreasonable (and undesirable). In practice, the steering committee agenda should be made up only of above-the-line agenda items, but participants should accept that discussion will inevitably "stray" below-the-line. It is up to the project owner (or whoever is acting as Chair) to ensure that the steering committee does not degenerate into an alternative project team.

A steering committee meeting agenda that is consistent with the previous reporting template is offered in Box 7.4: For each substantive agenda item, there are the usual three sub-items: report, discussion, decisions.

7.5.3 Celebrating Success

Life can be tough for project teams and it is very common for members to feel that their (often extraordinary) efforts go unrecognised. Item #2.3 ("deliveries") in the standing agenda for the steering committee offers an important opportunity to boost team morale. As it has been put so simply by many others "Celebrate success!". Every time the team delivers a significant output, its efforts should be acknowledged and, if appropriate, rewarded by the steering committee.

7.6 Other Key Players

During execution, reference groups and advisers will fill the roles defined in their brief. Much of this work will take the form of meetings, working sessions and the preparation of commissioned reports. Periodically they will meet with the project manager (or the project owner) to report on their work and review their overall effectiveness.

Project assurance counsellors undertake a wide range of tasks during execution, including:

- Conducting above-the-line reviews of project documentation.
- Meeting with the project manager, project owner and other key players as required.
- Preparing regular project assurance reports for presentation to the steering committee.

Project counsellors work in close consultation with the project manager so that there are never any surprises in reports tabled at a steering committee meeting. Counsellors are not auditors and so they are free to work in an advisory—as well as an assurance capacity. Both the project counsellors are involved in a regular cycle of activity throughout execution based on:

- Review of all substantive above-the-line document such as updates to the project plans and business case.
- Review of all regular reports as they are tabled for the steering committee. (It is not necessary for the counsellors to vet reports before they go to the steering committee, but, on occasions, that may be useful).
- Periodic participation in team meetings.
- Preparation of periodic reports to the steering committee.
- Participation in meetings of the steering committee. Care has to be exercised by both that they keep a relatively "low profile" in these meetings so that their active involvement is confined to discussion of their own reports and issues associated with their respective roles.
- Involvement in regular ad hoc meetings with the project owner, project manager and other key players.

7.7 Summary

We have examined the (extensive) above-the-line activity that is carried out during execution. At this point, the bulk of the work that defines the traditional model of a project is complete. Because of this, many key players will see the project as over and so they will quickly lose interest. They may even demand that "everything be wrapped up quickly", despite the fact that the original target outcomes have not yet been realised and the objective has not yet been achieved. This pressure must be resisted, so that outcome realisation can be undertaken as a critical final phase in the life of the project. That is the subject of the next Chapter.

Chapter 8
Realising Outcomes from a Project: The Roles of the Key Players

8.1 An Outline of Outcome Realisation

Outcome realisation is led by the project owner and begins with the early utilisation of outputs, after they have been delivered by the project manager. Outcome realisation is completed (and the project is closed) when the flow of target outcomes is secured. During this phase, the project owner seeks to support the project's customers in their utilisation of the project's outputs. Utilisation can take many forms, for example working to a new organisational procedure, viewing (by patrons) of paintings exhibited by the national museum, or promoting a product with new marketing collateral. In some projects, utilisation continues indefinitely into the future (after outcomes have been secured) by transferring outputs into the operational environment. Other projects close without any involvement of business operations at all. In cases where outputs will be used in an operational environment, following securing of the flow of target outcomes, outputs are handed over to business operations and the project is evaluated. Outcome realisation has two major activities, facilitation and outcome closeout, described in Fig. 8.1 and discussed in the following sections.

A project can be declared closed when its target outcomes are secured. Target outcomes are secured if any of three conditions are met:

1. They reach target thresholds and there is an acceptable probability that they will continue at this level into the future.
2. They fall short of target thresholds and there is no evidence that they will reach their thresholds in the foreseeable future.
3. No target outcomes are generated at all and there is no evidence that they will appear in the near future.

In those (unfortunate) cases where a project is aborted before all outputs were delivered, it will already have been closed in the previous phase (whenever execution was terminated).

O. Zwikael and J. Smyrk, *Project Management for the Creation of Organisational Value*, DOI: 10.1007/978-1-84996-516-3_8,
© Springer-Verlag London Limited 2011

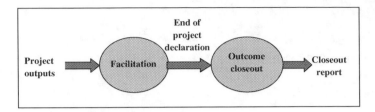

Fig. 8.1 The outcome realisation processes

8.1.1 Natural and Synthetic Outcome Realisation

It is necessary to briefly return to the concept of natural and synthetic outcomes (introduced in Chap. 2) before discussing the processes of outcome realisation. The generation of natural outcomes does not require that outputs be utilised by customers. This does not imply, however, that target outcomes will be achieved automatically. In the case of Ripple Rock (see Box 2.4) for example, it is conceivable that shipping may have been banned from the Seymour Narrows for other reasons, and so the end effect on shipwrecks attributable to the destruction of the Rock would have been zero.

For projects with natural outcomes, outcome realisation involves little more that "watching and waiting"—in other words, measuring the outcomes that were sought and waiting until a declaration can be made about their achievement. With synthetic outcomes (where outcomes are generated through the utilisation of outputs), this final phase involves (proactive) processes of facilitation and intervention. Whereas synthetic outcome projects are closed when the flow of target outcomes is *secured*, projects with natural outcomes are closed when the flow of target outcomes is *realised*. The choice of words here is significant: *secured* carries with it the idea of intervention, while *realised* suggests passive observation.

8.1.2 Facilitation

The project owner leads a group that supports utilisation of outputs by customers to ensure that target outcomes are being generated effectively. The project owner is accountable to the funder for securing target outcomes. The project owner may retain the project manager to administer this work, although he/she will not be accountable for the results from this phase. One of the major contributions to conventional wisdom from the approach presented in this book is the additional focus on outcome realisation by the project owner.

The need to facilitate output utilisation arises because, in many cases, customers struggle with the (frequently unwelcome) changes that the arrival of new outputs imposes. It has been found that project performance is sensitive to the capabilities of the facilitating team and the quality of the preparations undertaken for customers (Dvir 2005). In addition, during this phase it is also critical that the project owner continues to sell the project to the customers.

Deciding upon the correct amount of time to spend on facilitation before handover of responsibility to the line manager is not an easy decision. Too short a facilitation process may mean that customers do not become familiar with the new outputs and as a result, their utilisation may not be effective. Too long a facilitation process, on the other hand, may cost additional money and inhibit the line manager from assuming full responsibility for any ongoing operations.

8.1.3 Handover

Where a project involves ongoing use of outputs by a business unit, the project team hands over those outputs to an operations area. In other cases, there is no such handover process. Handover would typically involve:

1. The responsibility for the new process.
2. All outputs from the project.
3. Relevant documents and operations manuals related to these outputs.

The activities undertaken in outcome realisation will depend on who had served as project owner:

1. *The line manager as a project owner.* The project is declared closed when the flow of target outcomes is secured. At that point, the role of project owner simply comes to an end and the person concerned continues in the role of line manager (there is no true handover of responsibility for future outcome generation).
2. *Handover from a project owner to a line manager.* Where someone other than the line manager has been appointed as project owner, there is a formal handover of the operation from the project manager to the line manager.

The point of handover (when it occurs) represents the end of the project, which requires only a simple declaration to make it official.

8.1.4 Outcome Closeout

Outcome closeout is undertaken for similar reasons to those that underpin outputs closeout, but with a different set of performance areas. As before, the process is centred on workshop and results in a second report, which not only summarises some further learnings from the exercise, but also provides the information on which a judgement can be made about the success of project ownership and investment. In most cases, there will be only one outcomes closeout, however it is conceivable that, for a project involving outcomes with disparate lead times, more than one may be required.

The outcomes closeout workshop would normally be conducted by the project owner, although this task may well be delegated to the project assurance counsellor or a professional facilitator. It is important for members of the steering committee to participate for two reasons: they are best equipped to contribute to the process and their interest sends a strong signal to the organisation that outcomes are crucial to future projects. In addition, key members of the team and major reference groups should also be involved where this is feasible.

Reflecting its continuous improvement origins, the technique for outcome closeout involves the following steps:

1. *Decide on a list of performance areas.* These are aspects of the project where the key players believe there are important lessons for the future. It is here that the project assurance counsellor may be able to assist by drawing on any list of assurance issues that he/she has logged in the course of the exercise. A generic list of outcomes closure performance areas is suggested in Table 8.1. This has two columns: "Performance area" and "Focus question". Each performance area is the subject of analysis and judgement during the workshop. Each focus question is intended to provide a target for the analysis and judgement. Although the list provided here is reasonably extensive, it would be desirable to restrict the number of performance areas that are considered in each workshop, perhaps to around seven.

2. *Evaluate the gap between desired and actual performance.* This is done on a scale of 0 through 5 with: 0 representing no gap (perfect performance) and 5 representing a large gap (poor performance). In the following templates, the rungs on the left hand "ladder" indicate the amount of gap between desired and actual performance. The star is located on the ladder according to the perceived gap.

3. *Identify the evidence.* What factual observations (positive and negative) can be catalogued in support of the "gap score" that has been given (sometimes it will

Table 8.1 A generic list of outcome closure performance areas

Performance area	Focus question for evaluation of the performance area
Overall worth	What is the project's actual worth?
Fortuitous outcomes	What other flows of desirable outcomes were attributable to the project?
Target outcomes	Did we achieve our target outcome flows?
Outcome timeframe	Did we realise our outcomes by the target dates?
Undesirable outcomes	Did the undesirable outcomes from the project accord with expectations?
Detrimental outcomes	Were there any undesirable outcomes that could be regarded as unexpected, unacceptable and avoidable?
Scope (ex post)	Was the agreed scope right? Were the delivered outputs fit-for-purpose? Did we identify the right outputs in the scoping statement, and were they given the correct fitness-for-purpose features?
The steering committee	How well did the steering committee work, and how closely did it adhere to its charter?

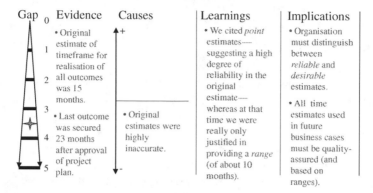

Performance area: Outcomes timeframe

"Did we realise our outcomes by the target dates?"

Gap	Evidence	Causes	Learnings	Implications
0	• Original estimate of timeframe for realisation of all outcomes was 15 months.	+	• We cited *point* estimates— suggesting a high degree of reliability in the original estimate—	• Organisation must distinguish between *reliable* and *desirable* estimates.
	• Last outcome was secured 23 months after approval of project plan.	• Original estimates were highly inaccurate.	whereas at that time we were really only justified in providing a *range* (of about 10 months).	• All time estimates used in future business cases must be quality-assured (and based on ranges).

Fig. 8.2 Sample of a report on a performance area examined in an outcome closeout workshop

be necessary to catalogue the evidence first and then talk it through until a consensus score can be agreed).

4. *Infer the causes of performance levels.* These are divided into two lists: the positive causes are those that gave rise to a good (that is, a small) gap, or that prevented the gap from getting even larger. Negative causes are those that gave rise to a poor (that is, a large) gap, or that prevented the gap from getting even smaller.

5. *Identify the learnings.* These take the form of statements that, if accepted as rules to guide the conduct of future projects, would generate better performance in this area in future.

6. *Discuss the implications.* What now needs to be done by those who were involved (individually and collectively) to "institutionalise" the learnings.

Figure 8.2 suggests a template for a report on each performance area that has been attacked in this way, showing typical entries for a hypothetical analysis of one outcome performance area.

8.1.5 Evaluation of Project Ownership

At the end of outcome realisation, but outside the closeout process, the funder should formally evaluate the performance of the project owner, based on the extent to which he/she achieved the approved business case. This should be undertaken in accordance with the organisation's human resources practices, but focused on achievement of the last approved business case.

If the project owner reports to the funder (in the line sense) then the results of the evaluation can be incorporated into annual assessment of the project owner.

If the project owner reports elsewhere in the organisation, then a report on the evaluation would be sent to the relevant executive for incorporation into the regular human resources assessment process.

8.2 The Project Manager

The project manager discharges his/her project accountabilities when all outputs have been delivered in accordance with whatever specifications have been agreed with the project owner. However, because the outcome realisation phase involves a significant amount of work, and in view of the project manager's intimate familiarity with the outputs that have been implemented, he/she is a leading candidate to support the project owner throughout this phase. While strictly speaking there is now no longer a project manager, it is inevitable that he/she will still be identified as "the project manager". It is important, however, that all key players understand that during this phase the role is purely one of support to the project owner. In what follows we outline this activity.

8.2.1 Project Customer Support

The assistance required by customers while utilising a project's outputs for the first time may include:

1. *Technical support.* Involving rectification of any problems with the project's outputs that have been uncovered during utilisation.
2. *Minor improvements to outputs.* Sometimes it will become clear that certain outputs, although delivered as specified, require minor modification to enhance their utilisation, often triggered by requests from key customers. Provided that the costs involved are acceptable to the funder, then this work may be carried out as part of outcome realisation. There will be thresholds of modification, beyond which the work would become the foundation for a follow-up project.
3. *On the job training.* New staff who join the organisation during the outcome realisation phase require training and induction which should also be carried out as part of this phase.

8.2.2 Utilisation Monitoring

At this time it is necessary to ensure that utilisation of each output is occurring as planned in the storyboard (introduced in Chap. 5). Table 8.2 shows how the storyboard can be used during outcome realisation, by comparing expected customer utilisation behaviour with actual experience (as is documented in the right hand column). In case of major differences, corrective action items are also suggested.

Table 8.2 Example of an actual story board for Shanghai tourism initiative

Project	Project customer	Output	Target outcome
Shanghai tourist promotion	Tourists from Shanghai	Plaques at points of interest	Increased visitation to the township
Storyboard			
Project customer utilisation step	Implied fitness-for-purpose features of plaques		Actual experience of project utilisation and related action items
Tourist sets out along heritage trail to next plaque	Clearly signposted from heritage trail Easily found from heritage trail		As anticipated
Stops at plaque to read	Dual English/Chinese Readable from outside		As anticipated. The Chinese translation was found to be very effective
Views features at point of interest	Accessible viewing points Brochure display stand Protected seating at some locations		Because of the long route, more protected seating should be added
On return to Shanghai, describes the experience to acquaintances, friends and relatives	Material displayed on plaques should be interesting, attractively displayed and noteworthy		Yet to be confirmed

8.3 The Project Owner

As a leader of this phase, the project owner is expected to secure the flow of target outcomes before a handover of any operational outputs to the line manager.

8.3.1 Ensuring Effective Utilisation of Outputs

Securing the flow of target outcomes requires that outputs are utilised effectively (according to expectations). The utilisation map (introduced in Chap. 5) can be employed as a monitoring tool during outcome realisation, as suggested by Table 8.3.

The analysis in the right hand column suggests changes whenever the achievement of target outcomes is in doubt. In the example, changes with formation of the supplier panel appear to be necessary in light of the analysis.

8.3.2 Outcome Evaluation

A comparison of actual (or predicted) results for each outcome with the agreed target can be conducted only when reliable data is available. Outcome measures can

Table 8.3 Actual utilisation map for Project BuyRite

Project output	Target outcome		Observations about actual utilisation and related action items
	Reduced procurement costs	Reduced payment times to our suppliers	
New procurement processes	Procurement department staff Suppliers	Procurement department staff	As anticipated
A panel of preferred suppliers	Procurement department staff	N/A	Although established, the panel has not yet met due to political disagreements with one supplier

only be made after a reasonable period of utilisation (which, in turn, requires that relevant outputs have been successfully delivered and implemented). In many cases, the target outcomes from a project are generated over an extended horizon into the future and so a declaration that they have been secured is, of necessity, based on both actual and forecast data. The actual observation confirms (or otherwise) that the target threshold has been reached (one of the tests that outcomes have been secured). Forecasts (together with the project's residual risk) will indicate the probability of meeting outcome targets into the future. As a result, there are three pieces of data involved in the judgement about each target outcome:

1. Agreed performance thresholds.
2. Measurement of actual values.
3. Calculated predictions of future values.

The comparison of actual and predicted data with the agreed target is conducted separately for each outcome. Such comparisons are made not only about the magnitudes of the variables involved, but also about their timings (as indicated in the following example).

Table 8.4 provides an illustrative comparison of this type in Project BuyRite for the outcome entitled "reduced procurement costs". This measure was expressed in the approved business case as the average cost (in dollars) per purchase order issued over a quarter (including consumables and labour). The conclusion from this table is that the target outcomes have exceeded expectations, while still delivered on time.

Table 8.4 Target and actual outcomes in Project BuyRite

Attribute	Target outcome title	
	Target outcome	Actual outcome
Target	A reduction of 25%	A reduction of 30%
Achievement date	Target to be realised by Q4 2011	Target was realised on time

The next section discusses the analysis of outcomes, the calculation of actual project worth and the judgement of ownership and investment success.

8.3.3 Project Evaluation

Actual outcomes play a central role in the evaluation of ownership and investment success. As mentioned in Chap. 3, ownership success is judged by the funder based on achievement of the business case, while investment success is based on the acceptability of the project's realised value. These judgements involve variations of a regression test:

1. Replace the values of certain parameters appearing in the approved business case with actual values to create the "realised" business case.
2. In the case of project ownership success, ask the funder: Do you accept the "realised" business case as equivalent to (or exceeding) the approved business case? If the answer is "yes", then project ownership is judged as successful. If, however, the answer is "no", then project ownership is judged as unsuccessful.
3. In the case of project investment success, ask the funder: If at the time of your earlier decision you had been given the realised business case, would you have approved it? If the answer is "yes", then project investment is judged as successful. If, however, the answer is "no", then project investment is judged as unsuccessful.

To conduct both of these tests, the funder will be particularly interested in the project's actual worth and any residual risk.

The actual worth of the project is calculated using three classes of variable:

1. Desirable outcomes (benefits).
2. Undesirable outcomes (disbenefits).
3. Cost.

For most projects there is normally a small number of variables in each class. For example, in project BuyRite there are two target outcomes. Each variable is measured and given a qualitative index representing its value to the funder. In the case of target outcomes, each was defined in the approved business case with a specified unit of measure and method of measurement. This provides the basis for the actual figures that appear against worth in the realised business case.

The residual risk for the project overall can be obtained by aggregating the risk exposures attached to all risks appearing in the risk register. The residual exposure for the project relates to any remaining likelihood that the project could still fail due to a catastrophic collapse in its future performance. In general, that sort of risk is associated with uncertainty about the benefits and costs attached to future business operations. In those projects where outcome generation is time-limited, the residual risk exposure is (by definition) zero.

Consider the following scenario for the example of the drug manufacturer introduced in Sect. 3.3.6. The project has experienced major cost overruns throughout its execution, exceeding the approved budget by 100%, brought about by two factors:

1. A number of significant improvements in the quality of the drug (which are expected to increase sales) have led to (approved) increases in expenditure.
2. Poor management of purchases associated with complex laboratory trials led to excessive outlays that caught management by surprise.

The analysis of the drug project's worth in Table 8.5 tells us two things:

1. Significant cost overruns (attributable to poor procurement management), caused the actual worth of the project to be lower than anticipated in the approved business case. Therefore we can conclude that this was both a project management failure (because of the cost overrun) and a project ownership failure (because of the significantly reduced project worth).
2. Despite the cost overruns, the project's actual worth is still positive. This strengthens the claim that the project can be considered an investment success, regardless of the judgements made about management and ownership success.

Figure 8.3 illustrates how judgements would be made about an arbitrary project "P". Consider the approved business case for "P". The worth and risk exposure for such a project indicate that it was worthy of funding (because it is located above the project investment frontier). We now examine three possible scenarios describing the project at its conclusion. In scenario "A", the realised business case exceeds the expectations attached to "P" and so would be regarded as a project ownership success. By way of contrast, in scenario "B", the realised business case falls into the region where, it falls short of the expectations attached to "P" and so would be regarded as a project ownership failure. It is worth emphasising that although "B" is associated with project ownership failure, it nevertheless, would

Table 8.5 The target and actual worth of a project

Element	Target values	Actual values
Benefits	$2,000M	$2,250M
	$400M per year	$450M per year
Disbenefits	$150M	$150M
	Lost revenue from cannibalisation of an existing product for a value of $30M a year	Lost revenue from cannibalisation of an existing product for a value of $30M a year
Cost	$665M	$1,205M
	Below-the-line costs = $500M	Below-the-line costs = $1,000M
	Above-the-line costs = $40M	Above-the-line costs = $80M
	Annual operations costs (to produce the new product) = $25M per year	Annual operations costs (to produce the new product) = $25M per year
	$500 + 40 + 25 \times 5 = 665$	$1,000 + 80 + 25 \times 5 = 1,205$
Project's worth	$1,185M	$895M
	(2,000-150–665)	(2,250-150–1,205)

Fig. 8.3 Evaluating project ownership and investment success

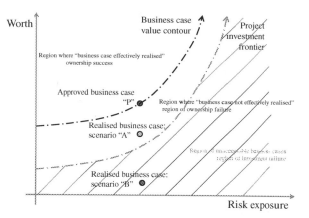

be evaluated as a project investment success (because it lies above the project investment frontier). Scenario "C", on the other hand is both an investment and ownership failure (because it falls below the project investment frontier).

8.4   The Steering Committee

The role of the steering committee during the outcome realisation phase is confined to monitoring and facilitating utilisation, to ensure that all the following happen:

1. *Support.* That all customers receive the required level of support so that they can transition smoothly into the new working environment (using the new outputs).
2. *Handover (in those cases when the project owner and the line manager are different entities).* That there is a complete and smooth transition from the project environment to the operational environment.
3. *Evaluation.* That the project is properly evaluated.
4. *Preparation for the future.* That appropriate steps have been taken to "institutionalise" the learnings from the outcomes closeout report, so that the organisation is committed to enhancing its future project performance.

8.5   Other Key Players

Other key players will either have stood down, or have relatively minor roles during this phase:

- *Reference groups and advisers.* Most of these will have stood down at the end of execution. In one or two instances, they may remain active during outcome realisation. If, for example a customer reference was established earlier, then it may well work through this phase, to participate in any support activity that is undertaken.

- *Project assurance counsellors.* They continue in a support role, especially as far as the project owner is concerned. The project assurance counsellor may be retained to facilitate the outcome closure workshop.
- *The project team.* Members begin to stand down after all outputs have been delivered successfully. A small group may, however, remain to fill a support role for project customers.

8.6 Summary

The project has now come to its end. Hopefully, the target outcomes anticipated by the funder are all being generated and the investment is proving fruitful. All participants return to their ongoing roles within their organisation, but not for long—the announcement of a new project is just around the corner.

Appendix A
An Integrated Glossary of Project Management Terms & Definitions

This glossary is intended to provide rigorous definitions of the terms and concepts used in the frameworks presented here. It is an *integrated* glossary—which means that: key terms are used in particular ways, there is a one-to-one relationship between these terms and the concepts to which they refer and technical terms used in each definition are themselves defined elsewhere in the glossary. When a word below is underlined, it means that a definition for that term exists in this glossary.

A definition and/or a description (depending on the nature of the term) is provided in the second column for each entry. To encourage consistency, certain definitions include a recommended convention (called a "word structure") for labelling specific instances of the associated term. The definition/description is elaborated, where appropriate, with some supporting discussion in the third column. Examples and illustrations for selected terms are provided in the fourth column. The examples have no particular context, but the illustrations are drawn from the International Concrete Operations Inc. (ICO's) BuyRite project case study.

Term	Definition / description	Discussion	Examples & illustrations
Above-the-line AtL	Related to the administrative work on a project associated with its planning, monitoring, management and closure.	Above-the-line is an adjectival expression. It is used to qualify both outputs and activities, and refers to the administrative/management side of the project. "Above-the-line", as a project management concept, should not be confused with the same term used in accounting, marketing and advertising. Refer also: below-the-line.	Develop the business case is an above-the-line activity. The business case itself is an above-the-line output.
Accepted (project) practice	A collection of procedures that are judged by the project management profession (especially the Project Management Institute) as meeting certain criteria—and hence accepted for use in the management of projects.	The Guide to the Project management Body of Knowledge, for example, includes a significant catalogue of accepted (project management) practices. These are frequently called "best practices".	Assembling a Work Breakdown Structure and developing a Gantt chart for a project are examples of accepted practices.
Achieved business case	See Business case (realised.		
Activity	A formalised project process that is made up of tasks. Also used loosely to mean any "block of work".	A stylised three-level Work Breakdown Structure is used for discussion purposes involving: phases, activities and tasks. Word structure: Activities are expressed as imperatives (command phrases). Refer also: process, work package.	In Project BuyRite, the work to reengineer the procurement process will involve a phase "Analyse the "As-Is" procurement process. An activity in such a phase could be "Measure the elapsed time for an execution of the "As-Is" process.

(continued)

(continued)

Term	Definition / description	Discussion	Examples & illustrations
Ad hoc tasks	Small "blocks of work" that require no formal planning or monitoring.	The optimal way of handling ad hoc tasks is to "make them up as you go along"—correcting errors by simply redoing the work involved. Ad hoc tasks are usually not included in the WBS.	Normally one would not plan the work involved in correcting the draft of a set of minutes from a meeting—instead treating it as an ad hoc task.
Administrator	See Project administrator.		
Adviser	A provider of specialist input to the project who is not supervised by the Project team	An adviser is the name given to a "reference group of one".	In Project BuyRite the project manager hired a project management consultant to serve as her adviser in the implementation of the Earned Value Methodology for project management control.
Alterant	An output taking the form of a changed (rather than a new) artefact.	It is important to note that although both alterants and outcomes are expressed as participial adjectives, they are distinct concepts. Alterants always involve artefacts ("repaired bridge"), while outcomes always involve end effects ("reduced incidence of crime"). Refer also: output.	
Appraisal (of a project)	Assessment of a project carried out before a project is approved. (This is called ex ante assessment).	The decisions to fund and approve a project are informed by the appraisal contained in the business case. Any decision to abandon or continue a project in progress will be based on a revised appraisal. Refer also: assessment, evaluation.	
Artefact	The immediate, direct result of a (human-directed) process		Amongst the (many) artefacts that will be recognised in the outputs from Project BuyRite are:

(continued)

(continued)

Term	Definition / description	Discussion	Examples & illustrations
	that has a physical representation—with measurable physical attributes.	The criteria contained in the definition are necessary conditions for something to be an artefact. Refer also: output.	A model of the procurement process (in the form of a flowchart). A document outlining the company's procurement policies. An organisational model (for the new procurement unit). A configured enabling applications system.
Assessment (of a project)	A process to measure the attractiveness of a project.	Takes two forms: Appraisal—on which approval to proceed with the project is based. Evaluation—on which the quality of the (earlier) decision to proceed is gauged.	
Assessment framework	A framework by which the results of assessment can be used to make judgements about success.	Based on four concepts: Assessment processes. Assessment targets. Assessment tests. Assessment triggers.	
Assessment target	The level at which a project is assessed.	There are three levels of project assessment: The investment in the project The project ownership process The project management process	
Assessment test	A test that allows judgements of success or failure to be made about a project.	A test is made up of: A specific set of performance variables. A measurement on each of those variables. A set of criteria. A rule showing how to use the resulting measures to make a judgement about the project.	

(continued)

(continued)

Term	Definition / description	Discussion	Examples & illustrations
Assessment trigger	An event that signals the need for a measurement of project-related performance.	The triggers for the processes related to project performance are: Project appraisal: the tabling of a business case. Evaluation of project management performance: delivery of particular outputs. Evaluation of overall project performance: securing of target outcomes.	
Assumption	A condition placed on a project which, if not met, would require significant revisions to the business case and confirmation of the funding decision.	Typically assumptions take the form of expectations or they relate to the values of project parameters.	Examples of valid assumptions: "The US$/Euro exchange rate will not change by more than 2.5% over the rest of the year". "The manufacturing division will second three people for six months to the project team". Examples of pointless assumptions: "The business case will be accepted". "The project will be successful".
Attribute	A defined/measurable characteristic	The word attribute is used here in a sense very similar to that adopted in data modelling. In registers, entities take the form of instances (which are represented by rows), while columns are associated with attributes.	A date for realisation is an attribute of a project's target outcomes. A latest finish date is an attribute of a project task.
Attractiveness (of a project)	An index that can be used to rank projects (in terms of their desirability). Attractiveness is obtained by trading-off worth against riskiness.	The attractiveness of a proposed project determines how it ranks against competing ventures. This ranking can then be used by a funder to decide on when it will be funded (if at all).	

(continued)

(continued)

Term	Definition / description	Discussion	Examples & illustrations
Baseline document	A document used to define, approve, monitor and evaluate a project.	A baseline document is a comprehensive model of a project. The two major baseline documents for a project are: the business case and the project plan.	
Below-the-line BtL	Related to the outputs identified in the project's scoping statement.	"Below-the-line" is an adjectival expression. It used to qualify both outputs and work. Refer also: above-the-line.	For "Project BuyRite": Assessing optional designs for the new procurement process is a below-the-line activity. The list of preferred suppliers is a below-the-line output.
Beneficiary	A (project) beneficiary is a stakeholder who the funder wants to see "better off" as a result of the project.	An entity who is better-off because of the project, but whose gain is of no interest to the funder is not a beneficiary (in the formal sense). Such a stakeholder is a (positive) impactee. Beneficiaries are not to be confused with a project's customers.	The company (ICO) itself is a beneficiary of "Project BuyRite"—experiencing a flow of value from any reduced operating costs.
Benefit	A notional flow of value (in the eyes of the funder) to a beneficiary—attributable to the achievement of desirable outcomes.	A particular benefit stream will qualify for inclusion in the assessment of the project only if the funder believes it to be an appropriate "return" on his/her investment. A benefit is derived as a qualitative index from measures of target outcomes.	In Project BuyRite, lowered operating costs are a benefit from the project—flowing to the company itself. The achievement of reduced supplier settlement times implies a benefit to vendors—in the form of reduced working capital. The notional value attached to this effect by ICO on one hand and the vendors on the other, may differ.
Budget	The total value of the pool of economic resources approved for the project.	The budget covers: Approved real outlays on purchased resources from outside the funding entity.	

(continued)

(continued)

Term	Definition / description	Discussion	Examples & illustrations
		The opportunity cost of all approved internal resources. Any costs associated with future operation of a project's outputs are not normally included in the project budget.	
Business case	A baseline document that provides all the information required to make a reliable decision about funding a project.	The business case is owned and tabled by the project owner, but usually assembled by an experienced project manager (as the project owner's agent). The business case serves as a project charter.	The business case for Project BuyRite will be assembled for Nancy Palmer by Paul Myer.
Business case (realised)	A notional version of the business case based on what actually transpired—rather than on what was intended to happen.	The realised business case contains only verifiable actual information. It is used to judge project success in the business case regression test.	
Business case (approved)	That version of the business case on which the last substantive decisions about the conduct of the project were based.	The approved business case is the last modified business case accepted by the owner.	
Business case (modified)	A revised and updated version of the (approved) business case.	The modified business case differs from the original in two respects:	

(continued)

(continued)

Term	Definition / description	Discussion	Examples & illustrations
		Selected parameters (such as timeframes) are revised.	
		Certain sections (such as the risk register) are moved into the project plan.	
Business case (approved)	The business case (produced during initiation) on which the initial approval for funding is based.	At the end of initiation a funding decision-in-principle is taken—based on the approved business case.	
Business environment	The environment within which operational processes are executed.	The business environment is distinguished from the project environment. In many projects, certain outputs are transferred to the operational environment for ongoing generation of target outcomes. Refer also: project environment.	
Business owner (of an output)	The operational entity accountable for any ongoing utilisation of one or more of a project's outputs after the project is declared closed. The business owner is the operational equivalent of a project owner.	Business ownership is a (core) role in the operational environment of a business. A business owner is a stakeholder in a project—but does not have a distinct role in the project environment. The role of business owner begins after the project is complete. In certain cases, the business owner of an output may be a candidate for the role of project owner. Refer also to two related defined terms in the ITO model: project owner, project customer	Nancy Palmer is the owner of Project BuyRite and, as National Procurement Manager, will (in due course) become the business owner of all the outputs that the project will eventually deliver (such as the Panel of Preferred Suppliers).
Business process	A regularised, significant operational "block	"Big" processes that have a formal operational model are called "business processes". "Small" processes for which an operational	Amongst the (many) processes which ICO executes in the course of doing business are: Quote customers.

(continued)

(continued)

Term	Definition / description	Discussion	Examples & illustrations
	of work" for which a formal model exists—drawn from a long history of repetition.	model exists are called "procedures". "Small" processes for which no operational model exists are called "ad hoc tasks". The scope of a business process is defined by its outputs. In contrast, projects are "big" processes for which no formal model exists (at least until the business case is assembled).	Procure supplies. Manufacture concrete. Deliver concrete. Handling a customer complaint is formalised in ICO—and hence is an example of a procedure. The work of planning the itinerary for plant visits by executives from Central Office is not formalised—and so is an example of an ad hoc task.
Champion	The person who seeks approval from the sponsoring entity to assemble (or who is assigned responsibility by the sponsoring entity for the assembly of) a business case.	The champion is the person or entity who drives the project through initiation to approval. The project champion is sometimes called the "promoter". The champion may become the Project owner.	National Procurement Manager (Nancy Palmer) is the champion for Project BuyRite. She also happens to becomes project owner.
Charter	A specification of the role to be played by a group of participants in a project (such as a Steering Committee).	Refer also: Role description, Terms of reference.	
Classes of output	See Output classes.		

(continued)

(continued)

Term	Definition / description	Discussion	Examples & illustrations
Closeout	A process of judging a completed (or abandoned) project across a number of selected performance areas—with a view to gauging performance and improving the organisation's capabilities for future initiatives.	A Closeout report is the output from the closeout process. Closeout is distinct from Post Implementation Review (PIR)—a formal technique to determine if an output is fit-for-purpose. There are two classes of closure: Outputs closeout: undertaken when selected outputs are have been delivered—or the project abandoned. Outcomes closeout: undertaken when outcome flows have been secured.	
Closeout report	A report on the performance of the project across a number of selected performance areas (such as adherence to timeframes).	The closeout report gauges the success of the project and lays the foundation for future project performance improvement. In a large project there may a succession of closeout reports.	
Commissioned stakeholding	A form of stakeholding arising when an entity is appointed to fill a role in the governance model of a project.	There are nine generic classes of commissioned stakeholder in a project: Advisers. Champion. Counsellors. Project Manager. Project Owner. Reference group members.	Paul Myer is a commissioned stakeholder in Project BuyRite.

(continued)

(continued)

Term	Definition / description	Discussion	Examples & illustrations
Communication plan	A comprehensive framework to guide development and delivery of the communications that will be effected throughout the project.	Steering committee members Suppliers. Team members. Refer also: Spontaneous stakeholding. One of the three generic forms of stakeholder engagement is inclusion in a communications plan. A communications plan integrates all such forms of engagement for all stakeholders.	
Constraint	A bound that is set by key players or imposed by circumstances on a project parameter.	Constraints are usually set for: Timeframe. Cost (outlays/resources). There is no guarantee that a meaningful project can be scoped within all the constraints that are placed on it and so, in some cases it proves necessary to violate the constraints that have been set for these two parameters. Other constraints that are imposed may be absolutely inviolable.	"Only$2.5 M in cash is available for the project". "We are required to comply with the new Companies Law by the end of August". "The site is inaccessible during summer because of the monsoon"
Contingency	A risk mitigating action that will have the effect of reducing the Severity (of the damage) to a project	Refer also: preemptive, infeasible project	Consider a threat to a project—the project manager leaves. A candidate contingency action for this threat might be "train a team member as deputy project manager".

(continued)

(continued)

Term	Definition / description	Discussion	Examples & illustrations
	arising from the realisation of a particular threat.		
Counsellor	See Project counsellor		
Contingency fund	A portion of the project budget set aside for either: Undertaking any necessary contingency actions (as part of the project's risk mitigation programme). Unforeseen risks.	A contingency fund is not a "slush fund"—all draw-downs must be justified and documented.	
Cost(s)	The economic value of resources demanded by the project and the ongoing operation of its outputs.	Costs arise from outlays of cash (to acquire products/services) and the deployment of internal staff. Costs are of two kinds: Those incurred during the project itself (which may be further divided into management (AtL) and production (BtL) costs). Those required for the ongoing operation of the project's outputs. Costs and disbenefits are distinct concepts.	

(continued)

(continued)

Term	Definition / description	Discussion	Examples & illustrations
Cost-infeasible (project)	A project for which the resource plan indicates a demand for more resources than those approved in the budget. The difference between the budget and the costs of the resource plan is a gauge of the level of (cost) infeasibility.	Cost infeasibility must be resolved before a project can be approved. Refer also infeasible project.	If Charles Edwards indicated a budget of $50M for the exercise—but Paul Myer's initial estimate is $60M—then the project may well be infeasible and must not be approved until the infeasibility is resolved.
Counsellor	See Project counsellor		
Criterion (singular) criteria (plural)	A threshold of performance taking the form of a value for a particular performance variable.	Refer also: performance variable.	
Customer	See Project customer.		
Customer map	See Utilisation map.		
Deadline	An arbitrarily set date for an event in the project.	Not to be confused with milestone. A deadline has no supporting analysis confirming its achievability. A project cannot be scheduled/tracked to deadlines—only to milestones.	
Defining (a project's) outputs	A project's outputs are defined when two conditions have been		Core outputs for Project BuyRite include: A new procurement process (represented by flowcharts and supporting descriptions).

(continued)

(continued)

Term	Definition / description	Discussion	Examples & illustrations
	met: All the outputs from the project have been identified and catalogued. Critical fit-for-purpose features have been set for each.	In practice, the cataloguing of a project's outputs might require a PBS (product breakdown structure)—rather than a simple list. Refer also: setting (project) scope, scope	A panel of preferred suppliers. A suite of enabling applications systems. A new organisational model.
Deliver (a project output)	To transfer an output from the project environment to a project customer—in readiness for implementation.	Delivery is the second of three steps associated with creation of an output (the other two being production and implementation). The accountability of the project manager is discharged on delivery. Refer also: produce, implement.	
Deliverable	An output that appears in a statement of scope.	Deliverables are final outputs—as distinct from intermediate and above-the-line outputs. Final outputs are necessary for project success (and also used to judge project management success).	In project BuyRite, the new (to-be) procurement process is a deliverable. A flowchart describing the old (as-is) processes is an intermediate output, while a Gantt chart appearing in the project plan is an above-the-line output.
Dependency	A linkage amongst elements of a Work Breakdown Structure indicating a constraint on their timing.	In the forward scheduling algorithms for workplanning there are three forms of dependency for pairs of activities ("A" & "B"): S-S (start to start). F-F (finish to finish). F-S (finish to start).	

(continued)

(continued)

Term	Definition / description	Discussion	Examples & illustrations
Descope (a project)	To reduce the scope of a project.	A project can be descoped in either/both of two ways: Remove outputs from the scoping statement. Remove (or weaken) the fitness-for-purpose features of selected outputs.	
Desirable outcome	An outcome that has value in the eyes of the project funder	Refer also: outcome.	
Detrimental outcome	An undesirable outcome that is also unexpected, unacceptable and avoidable.	Detrimental outcomes that are attributable to the actions of the project manager are taken into account when evaluating project management performance. (There may also be detrimental outcomes that have nothing to do with the project manager).	
Disbeneficiary	A project stakeholder who experiences any loss of value from the project.	A project's disbeneficiary is also a (negative) impactee.	Those suppliers to ICO who do not qualify for the Panel of preferred suppliers are disbeneficiaries of Project BuyRite.
Disbenefit	An expected notional loss of value by an entity (attributable to the project).	A particular disbenefit stream may or may not qualify for inclusion in the calculation of a project's worth.	The reduced volume of business from ICO suffered by those established suppliers who do not make it onto the preferred vendor list represents a disbenefit to them.
Division (of a project governance model)	A project governance model has four divisions (each related to the major roles of a project).	The four divisions are: steering, delivery, reference and assurance.	

(continued)

(continued)

Term	Definition / description	Discussion	Examples & illustrations
Duration	The time taken to execute a process—in particular any subset of a Work Breakdown Structure (including: a task, an activity or a phase).	Elapsed time and labour are closely related—but separately defined concepts	The unit of measurement for duration is typically days, weeks or months. By way of contrast: the unit of measurement for labour is typically person-days, person-weeks or person-months.
Earned value	A function in which the dependent variable is cumulative planned project costs and the independent variable is the schedule of actual achieved milestones.	This is derived from an extension to a schedule of milestones relating (for each) a progressive total of project planned costs to the actual (or forecast) date of completion. Refer also: planned costs	
Evaluation	Ex post assessment of a project—carried out after a project is executed.	Evaluation allows a judgement to be made about whether or not the original funding decision is cause for regret. Refer also: appraisal, closeout.	
Event	An identifiable instant of time.	Events can be: Internal—associated with the end of a task in the Work Breakdown Structure. External—associated with activity that lies outside the project's workplan. The ends of tasks in a WBS are events. Selected events are designated as milestones.	In Project BuyRite: The completion of the drafting of a new supplier contract is an internal event. The implementation of changes to related taxation law is an external event in Project BuyRite.

(continued)

(continued)

Term	Definition / description	Discussion	Examples & illustrations
Event-impact model (of a risk)	The event-impact model of a risk has three (sequential) components: A triggering event (threat). A chain of consequences. A (damaging) impact.	One instance of an event impact model is called "a risk". A risk has a property called Risk Exposure (RE) which is a measure of its importance obtained analytically from Li (likelihood) and Se (severity).	
Evidence of achievability	Reproducible, analytical evidence that confirms the achievability of estimates for timeframes and costs.	The reliability of schedules and budgets is gauged by the availability of such evidence.	
Expectation	A belief by a stakeholder concerning some parameter of a project	Risks to a project can arise from gaps between stakeholder expectations and the parameters established in the project's baseline documents.	
Execution (project phase)	A global phase of a project directed at creation of the project's outputs.	Execution is led by the project manager (accountable to the project owner). Execution begins when a funder approves a project plans. It ends when all of the outputs in the statement of scope are produced, delivered and implemented. (Or when a project is abandoned).	
Execution (of a process)	Undertaking a process.	Execution of a process is complete when the desired outputs have been delivered.	

(continued)

(continued)

Term	Definition / description	Discussion	Examples & illustrations
Failure	A state of unacceptable performance.	Refer also: performance, success.	
Fitness-for-purpose	Suitability for utilisation by a project customer.	Fitness-for-purpose is a binary concept related only to a project's outputs. Outputs are either fit-for-purpose or they are not. The term has two senses: Ex ante, "Does an output meet its agreed specification?" Ex post, "Was the output specified correctly?". The quality of an output is defined by its fitness-for-purpose features.	
Fitness-for-purpose feature	A characteristic of an output that makes it suitable for utilisation.	A critical property of an output from a project. The set of fitness-for-purpose features of an output provides the foundation for its eventual specification.	A requirement that in the new procurement process that suppliers will be paid when good are delivered (rather than when an invoice is received) is an example of a fitness-for-purpose feature (of the new process).
Flow of value	The stream of benefits (or disbenefits) that a stakeholder experiences (as valued by the funder).	Flows of value can be either: Positive—giving rise to a gain in value (benefit); or, Negative—giving rise to a loss of value (disbenefit).	The reduced procurement costs arising from Project BuyRite represent a flow of value to ICO. The faster payments for deliveries represents a flow of value for ICO's vendors.
Forms of stakeholding	The stakeholding that an entity has in a project can be of two forms: Spontaneous (arising from the nature of	There are six classes of spontaneous stakeholding in a project and nine classes of commissioned stakeholding (defined by the project governance model).	

(continued)

(continued)

Term	Definition / description	Discussion	Examples & illustrations
	the project). Commissioned (arising because of an appointment to fill an assigned role in the project).		
Fortuitous outcomes	Desirable outcomes that are not targeted in a project's business case.	Fortuitous outcomes can play no part in the (ex ante) appraisal of a project. Realised fortuitous outcomes are, however, recognised in project evaluation when gauging the project's actual (ex post) worth.	
Funder	A stakeholder who commits funds and/ or approves the allocation of labour to the project.	A project can have multiple funders. Note that the funder decides on funding The issue of who owns the money, is irrelevant to the definition.	In this case Charles Edwards (as CEO) has assumed the role of funder. (Although the shareholders "own" the money, they are not the "funders").
Funding organisation	The organisation to whom the funder belongs.	Refer also: performing organisation	
Gantt chart	A diagrammatic representation of a project schedule	In a Gantt chart: The vertical axis identifies elements of the Work Breakdown Structure. The horizontal axis shows time (as a calendar). Each element of the Work Breakdown Structure is represented by a bar indicating the start, duration and finish. Links between bars indicate precedence	

(continued)

(continued)

Term	Definition / description	Discussion	Examples & illustrations
Global phase (of a project) model	The life of a project is made up of four global phases: Initiation. Planning. Execution. Outcome realisation.	A gobal phase can be undertaken only of the one before has been completed. A project can be terminated at any point.	
Governance model	An organisational model that recognises and integrates responsibilities for outcomes, outputs and work by defining, for a structure of entities, their roles and relationships.	The governance arrangements for all organisations will normally recognise three: An overarching corporate (or institutional) governance model. An operational governance model. A (collection) of project governance models.	
Grading Table	A two-way look-up table that converts word values for Li (Likelihood) and Se (Severity) into a qualitative gauge of RE (Risk Exposure).	A grading table offers a way of "multiplying two words together" to obtain a "product" (as another word)	An example of a Grading Table follows—based on the two wordscales offered as illustrations elsewhere in this glossary for Likelihood (Li) and Severity (Se):

Likelihood	Hi	Med	Lo
Severity			
Grave	A	B	C
Moderate	C	D	E
Negligible	E	F	G

The letters "A" through "G" indicate (as an ordinal scale) the RE associated with the threat being analysed.

(continued)

(continued)

Term	Definition / description	Discussion	Examples & illustrations
Grading Rule	A pseudo-numerical procedure that converts qualitative values for Li (Likelihood) and Se (Severity) into a qualitative gauge of RE (Risk Exposure).	A grading table offers a way of "multiplying two words together" to obtain a product (as another word)	Consider the two simple wordscales used in the previous example of a Grading Table, but augmented with numbers so that: Hi has the value 3, "Med" 2 & "Lo" 1. A grading rule is: RE = (the number associated with Li) * (the number associated with Se). For example, if a particular threat has Li = Hi (3) and Se = Med (2), then its RE = 6.
Hierarchical decomposition	The recursive process of describing an element by breaking it into smaller elements.	Effectively, hierarchical decomposition converts a small number of large concepts into a large number of small concepts.	The most familiar model of a project obtained from hierarchical decomposition is the Work Breakdown Structure (WBS).
Impact	The last (of the three) components of a risk. An Impact takes the form of damaging effects that are reflected in a lowering of the worth of a project.	An impact has a property called severity (Se). Se is analogous to level of damage (or "loss of worth"). There are only three variables that determine the worth of a project—each of which can be "damaged" in two ways. Thus there are only six forms of damaging effect that a project can suffer: Consider the threat to Project BuyRite: "Software vendor experiences cashflow problems". In that case, two impacts would be expected: Benefits delayed, and Costs increased (both contributing to reduced worth for the project.

Parameter Variable	Timing	Magnitude
Benefits	Delayed (late)	Decreased
Costs	Advanced (early)	Increased
Disbenefits	Advanced (early)	Increased

(continued)

(continued)

Term	Definition / description	Discussion	Examples & illustrations
Impactee	A (project) impactee is a stakeholder who experiences either: A loss of value as a consequence of the project being undertaken—regardless of whether or not it achieves its target outcomes. (Negative impactees). A gain in value from non-target outcomes. (Positive impactees).	The interests of negative impactees are acknowledged in the business case and managed as a critical element of a project for either of two reasons: From an equity point-of-view, negative project outcomes should be managed. This can also be viewed as an issue of altruism. From a risk point-of-view, an impactee can become a serious threat to success. The interests of positive impactees are ignored in the business case—but could well figure in a project's final evaluation.	Those established suppliers who don't make it onto the preferred vendor list are (negative) impactees of Project BuyRite—because they will suffer a reduced volume of business from ICO. These firms are stakeholders in the project—and are to be included in the stakeholder engagement plan.
Implement (an output)	The activity of preparing an output for utilisation by a project customer.	Implementation is the last of three steps associated with creation of an output (the other two being production and delivery). Refer also: produce, deliver.	Implementation of the new procurement process in Project BuyRite could involve a trial period based on "live" purchases for ICO.
Infeasible (project)	A project for which parameters have been set that are internally inconsistent. There is no programme of work that would allow the parameters to be achieved—and so the project cannot succeed.	Infeasibility arises from constraints on either/both of: Costs/outlays/resources. Timeframe Refer also: cost-infeasible (project), time-infeasible (project).	

(continued)

(continued)

Term	Definition / description	Discussion	Examples & illustrations
Influencer	A (project) influencer is a stakeholder who, by virtue of his/her role, standing or influence is able to carry a significant body of opinion about the project.	Influencers are important targets of the project's communications plan because their views determine the views of a large number of stakeholders.	In Project BuyRite, Lindsay Thomas (a purchasing manager who has been with the company for 35 years) is considered an influencer because he is well-regarded by suppliers, staff and the Board.
Initiation team	The team commissioned by the project champion to assemble the business case.	Initiation teams tend to have a rather ad hoc structure—with members usually assigned for relatively short periods of time to work on specific sections of the business case.	
Initiation	A global phase of a project directed at assembly of a business case.	Initiation is led by a promoter (or a champion). This is someone who is normally quite senior—and who could well be a candidate for the role of Project Owner. Initiation begins when a funder requests (or agrees to receive) a business case. It ends when a decision is taken on funding the project. Initiation is the 1st phase of every project.	
Input	An economic resource that is consumed during the execution of a process.	In projects we are concerned with just two broad inputs: Labour. Purchased resources.	In Project BuyRite" significant inputs include: the time of assigned staff, the services of business improvement consultants and the costs of new applications system software.
Input-Process-Output (model) IPO	A (long established) model that seeks to explain the		

(continued)

(continued)

Term	Definition / description	Discussion	Examples & illustrations
	relationships amongst: inputs, a process and outputs.	The conventional view of a project corresponds to an IPO model of that project. The IPO model says nothing about outcomes (as defined in this Glossary).	
Input-Transform-Outcome (model) ITO	A model that seeks to explain the relationships between outputs and outcomes.	The ITO model was developed by John Smyrk in the late 80 s. It extends the IPO model by incorporating a mechanism called utilisation.	
Instance	An entry in a project register.	Project registers take the form of tables and so an instance is one "row" of entries.	For example an instance of risk (a single risk) is represented by one row of the Risk Register.
IPO	See Input Process Output.		
Issue	A matter of general concern for the project (other than a risk) that requires resolution.	An issue is distinct from a risk, but can evolve into a risk. The approach to issues and risk used here does not identify a realised risk as an "issue".	The definitions of issues and risks don't lend themselves particularly well to abstract discussion—examples tend to be more useful: "Fred Nurke has switched his rostered day off from Friday to Monday" is an example of an issue. "The project manager leaves" is an example of a risk (or, more precisely, a threat).
"Iron triangle"	A conventional test of success for project management.	The three conditions embedded in the "iron triangle" are that all the project's outputs be delivered:	

(continued)

(continued)

Term	Definition / description	Discussion	Examples & illustrations
	(Sometimes called the "golden" triangle).	To an agreed specification. On time. Within budget The "iron triangle" is neither necessary nor sufficient for project management success and there it cannot be used to draw any inferences about success or failure. Refer also: "seel tetrahedron".	
Issues management	The broad meaning: A (clerical) procedure for identifying, analysing, tracking and resolving issues.	The narrow meaning: The work of resolving an issue.	
Issues life-cycle	The particular sequence of states through which an issue passes between being identified and resolved.	The states that an issue can assume in its life-cycle are: To be assessed, Active, Inactive, Open, Parked, Resolved.	
Issues register	A tool for cataloguing issues and documenting their management.		
Issues report	A report that highlights important issues for consideration by a project Steering Committee.	The items appearing in an issues report are a subset of those identified in the issues register.	

(continued)

(continued)

Term	Definition / description	Discussion	Examples & illustrations
ITO	See Input Transform Outcome.		
Justification	"Project justification" is a practice accepted in some conventional project management methodologies by which the decision about funding a project is taken before the project is appraised.	There is often confusion in conventional thinking between "justifying a (committed) project" and "justifying the decision to fund a (committed) project". The former process is fundamentally flawed, but the second is meaningful.	
Key stakeholders	Those stakeholders in the project who are included in the stakeholder engagement plan.	Refer also: stakeholders.	
Likelihood Li	A qualitative measure of the level of confidence that a threat will occur.	Likelihood is similar to—but distinct from the concept of probability. Instead of a number between 0 and 1, Li is gauged using a word from a pre-fabricated wordscale (such as: Hi, Med, Lo). The Li of a threat before an RMP is put into place is called "Pre-Li". The Li of a threat after an RMP is put into place is called "Post-Li".	

(continued)

(continued)

Term	Definition / description	Discussion	Examples & illustrations
Loss of value	See flow of value.		
Measurability	A characteristic of an entity whereby at least one of its attributes can be gauged (qualitatively or quantitatively).	An output can have measurable attributes (such as mass, colour, readability). Outcomes must have measures if they are to be targeted.	
Memorandum of Understanding MoU	A formalised agreement under which an internal entity releases resources to a project.	MoUs lay out the terms under which the resource is made available and are somewhat akin to contracts—but have no legal standing.	
Methodology	A systematic, standardised high level model of a generic process. Such a model is used as a starting point in the development of a comprehensive script for a particular application of that process.	This term is used in two distinct ways in the project management literature: To indicate a recognised process (a method). For example there are established methodologies for undertaking social surveys. To indicate an entire framework of tools, techniques and templates for managing projects. Refer also: project management framework.	The systems development life cycle is an example of the word " methodology" used in the first sense. Software developers have standardised structures for the work involved in developing application systems. The Tasmanian Government Project Management Guidelines and PRINCE2 are both examples of the word "methodology" used in the second sense. Both are comprehensive frameworks of tools, techniques and processes to guide the management of projects.
Milestone	An event that has been adopted for use in tracking a project's	Milestones are derived analytically from the project's workplan and are not to be confused with deadlines (which are set arbitrarily).	

(continued)

(continued)

Term	Definition / description	Discussion	Examples & illustrations
	progress. These are usually selected from those events that define the finish of a task in the WBS.	Word structure: Milestones are formally expressed with a trailing past participle (such as "process design approved"). Since tasks in an existing WBS are expressed as imperatives, it is acceptable to adopt the name of a task ("approve process design").	
Modified business case	See Business case (modified).		
Negative impactee	See Impactee.		
Network diagram	A diagrammatic representation of project activities included in a Work Breakdown Structure, their durations and inter-dependencies using nodes and arrows.	Such a diagram is known in mathematics as a "directed graph".	In project management there are two forms of network diagram: Activity-on-arrow where: The arrows represent activities. The nodes indicate precedence amongst the activities. Activity-on-node where: The nodes represent activities. The arrows indicate precedence amongst the activities.
Objective statement	A succinct statement that explains why the project is being undertaken.	The (project) objective statement is one of three elements that make up a project's statement of scope (together with outcomes and outputs).	
Operational environment	That part of the business environment within which regular (non-project) business activity is conducted.	The two major classes of non-project work are: Ad hoc tasks. Transactions.	

(continued)

(continued)

Term	Definition / description	Discussion	Examples & illustrations
Original business case	See Business case (approved).		
Outcome	An indirect result attributable to an identifiable process or mechanism that takes the form of a measurable change in some variable that describes the "state of the world".	Outcomes are of two kinds: Desirable. Undesirable. Word structure: Outcomes usually begin with "Reduced ..." or "Increased ...". This rule is normally relaxed when dealing with binary outcomes—such as "Compliance with ...".	
Outcome realisation (project phase)	A global phase of a project directed at securing the flow of target outcomes.	Outcome realisation is led by the project owner (with support from the person who was project manager). Outcome realisation begins when customers begin utilising the project's outputs. It ends when the flow of target outcomes has been secured.	
Outlay(s)	Expenditure of cash on external products/services demanded by the project's Work Breakdown Structure.		
Output	A direct result from a process having a tangible physical representation.	In project management we are concerned only with outputs from human-directed processes. Such outputs are called artefacts and take either of two forms: a new artefact—or a change to an existing artefact (called an "alterant").	Primary outputs from Project BuyRite will include: A documented new procurement processes A new organisational model (for the Procurement/purchasing department)

(continued)

(continued)

Term	Definition / description	Discussion	Examples & illustrations
		Outputs are tangible. Outputs can be intermediate—or final. Final outputs are called deliverables. Word structure: Outputs are expressed as nouns. Verbs are not used—nor is the word "to".	New vendor contracts New application systems (IS) New technical infrastructure (IT).
Output classes	Outputs are classified in a typology according to the reason they have been included in the scope of the project.	1. ITO outputs: required for the generation of target outcomes. 2. Risk mitigation outputs: required for risk mitigation. 3. Stakeholder engagement outputs: required to engage stakeholder. 4. Mandatory outputs: required by law or policy. 5. Dependent outputs: required by another output or project.	
Output types	Outputs are of two types: intermediate, final.	Final outputs belong to one or more of the output classes and are recognised when judging project management success. Intermediate outputs are of no interest after final outputs are delivered.	
Overscoping	A situation in which the current scope of a project includes redundant outputs (or some outputs with superfluous fitness-for-purpose features)	An overscoped project incurs more costs and takes more time than its target outcomes require. Overscoping is commonly called "over-engineering". Refer also: underscoping.	

(continued)

(continued)

Term	Definition / description	Discussion	Examples & illustrations
Parameter	An attribute of a project that can be set or derived from other parameters. The parameters of a project include: Target outcomes/benefits. Scope. Risk exposure. Timeframe. Cost (outlays/resources).	There are only certain allowable subsets of the parameters for which values can be set arbitrarily.	Once the project has been scoped (so that target outcomes and outputs are internally consistent), then there are only certain combinations of values that are allowable for the other three parameters: Risk exposure. Timeframe. Cost (outlays/resources).
Performance	A gauge of achievement.	Performance is measured (or gauged). By way of contrast, success is judged (or determined)	
Performance variable	A variable used to gauge performance (on a project).		
Performing organisation	The organisation who engages/remunerates the project manager.	Refer also: funding organisation	
Phase	The highest level of the stylised three-level work breakdown structure discussed in the text. Phases are typically sequential.	For certain well-understood types of outputs (such as bridges, business processes and software) a standardised sequence of phases has been adopted as "accepted practice". Such a phase-sequence is called a "methodology"—one of a number of different meanings attached to this word).	

(continued)

(continued)

Term	Definition / description	Discussion	Examples & illustrations
Planning (project phase)	A global phase of a project directed at preparing a detailed model of the work required during execution.	Planning is led by the project owner (but undertaken by the project manager). Execution begins when a funder accepts a business case. It ends when a plan is tabled for consideration by the funder.	
Positive impactee	See Impactee.		
Predecessor	An element of a Work Breakdown Structure on that constrains in time the execution of another element.	Refer also: dependency, successor	
Preemptive	A risk mitigating action that will have the effect of reducing the likelihood of a threat emerging.	Refer also: contingency	Consider a threat to a project—the project manager leaves. A candidate preemptive for this threat is "formally contract the project manager".
Prioritise	To decide on the order in which independent: Outputs will be produced. Tasks will be executed.	In some conventional approaches to project management, the term is also used to suggest (incorrectly) that: Some outputs are optional—and thus can be "prioritised". Some work is optional—and can also be "prioritised".	ICO may well make a decision about the order in which purchased supplies are to be analysed and reengineered—and reflect that decision in a list of priorities. It would make no sense, however, to grade the outputs from Project BuyRite with

(continued)

(continued)

Term	Definition / description	Discussion	Examples & illustrations
		Using the theoretical principles discussed here we can conclude that neither of the above propositions is meaningful.	a priority list—because all outputs have to be produced (they wouldn't be in scope if that were not true).
Priority	See prioritise.		
Probity counsellor	A specialist independent advisor appointed by the SC to ensure that the commercial arrangements between the project team and the outside world are conducted in accordance with accepted procurement practice.	Otherwise known as a probity advisor. There is a large body of knowledge and accepted practice concerning the role of the probity counsellor.	
Procedure	A formalised "small" process of any kind.	Typically a formalised process becomes a procedure when it is recognised within some organisational context. Refer also: process, business process.	
Process	Human-directed work that produces one or more outputs.	Management in general and project management in particular is concerned only with human-directed processes.	All elements of a project's Work Breakdown Structure (for example: tasks, activities and phases) are processes—(as are all subsets of the elements that make up a Work Breakdown Structure). Procedures, practices and ad hoc tasks are processes.

(continued)

(continued)

Term	Definition / description	Discussion	Examples & illustrations
Produce (a project output)	The activity of assembling an output ready for delivery to a project customer.	Production is the first of three steps associated with creation of an output (the other two being delivery and implementation). Refer also: deliver, implement.	Implementation of the new procurement process in Project BuyRite could involve a trial period based on "live" purchases for ICO.
Product	A synonym for output.		
Programme (of projects)	A set of coordinated projects.	Projects may be coordinated under either of two different scenarios: One or other of the projects will suffer a reduction in its worth because one project "interferes" with the other. The current worth of one or other of the projects can be increased by electing to coordinate them.	
Programme of work	See workplan.		
Project	A unique process intended to achieve target outcomes.	Not all processes are suited to being managed using a project management framework. In general, processes that are novel and complex are best managed as projects.	
Project administrator	A team member who carries out selected above-the-line activities.	A project administrator reports to the project manager.	Examples of a project administrator's responsibilities: Manage the risk register. Prepare minutes of meetings. Confirm achievement of milestones.
Project counsellor	A specialist independent adviser appointed by the SC to ensure that specific aspects of	Counsellors are only appointed on large or sensitive projects—where there are significant governance, management, procurement, or quality issues.	

(continued)

(continued)

Term	Definition / description	Discussion	Examples & illustrations
	the project meet agreed standards. There are two forms of project counsellor: Project assurance counsellor. Probity counsellor.		
Project assurance counsellor	A specialist independent adviser appointed by the SC to ensure that the project is being conducted in accordance with accepted practice.	The Project assurance counsellor is not an auditor—and so is free to work consultatively with the project manager and other key players.	
Project baseline revision	The process used to change parameters set in the project's baseline documents.	Project baseline revisions are applied in particular to: The (current) modified business case. The current approved project plan.	Once accepted, revisions can be formalised as either: (cumulative) amendments to the relevant baseline document, or Revised versions of the relevant baseline document.
Project customer	An entity who, by utilising outputs, contributes to the generation of target outcomes.	A project can have multiple customers. The project customer is also called 'user' or 'end-user' (particularly in the IT sector).	In Project BuyRite, staff of the procurement unit are project customers (because they utilise the project's outputs to generate target outcomes—but they are not beneficiaries (in that they do not receive a flow of value from achievement of target outcomes).

(continued)

(continued)

Term	Definition / description	Discussion	Examples & illustrations
Project environment	That part of the business environment within which projects are initiated, executed and closed	The business environment can be viewed as having two parts: The project environment. The operational environment. Refer also: business environment, operational environment.	
Project environment management	The process used to manage external factors that are imposed on the project	Project environment management is applied in particular to: Stakeholders. Issues. Risks.	Registers are the primary tools in project environment management
Project execution control	The process used to identify and manage deviations from plan during the execution phase of a project. The above-the-line work surrounding execution	Project execution control is applied in particular to: Timeframes. Resources (outlays and/or labour). Output quality.	
		Project execution management involves three kinds of process: Project environment management. Project execution control. Project baseline revision.	Two forums are used to guide project execution management: Meetings of the project team (nominally weekly). Meetings of the steering committee (nominally monthly).
Project governance model PGM	A formal organisational model that: Identifies all the stakeholders who	A project governance model is separate from an organisational governance model—and, as a result, many of those involved in the project face a matrix management structure.	

(continued)

(continued)

Term	Definition / description	Discussion	Examples & illustrations
	will play a part in the project. Establishes the relationships and links amongst those stakeholders. Defines the roles of those stakeholders.	A PGM is linked into all the participating organisations through the line reporting arrangements of the individuals who will play a part in the project.	
Project management framework	A set of integrated, cohesive and related tools, procedures and techniques that can be used to guide the execution of a process.	Such a framework is useful in the management of processes that are large and novel. Refer also: methodology.	The Tasmanian Government Project Management Guidelines and PRINCE2 are both examples of the word "methodology" used in the second sense. Both are comprehensive frameworks of tools, techniques and processes to guide the management of projects.
Project Management Office PMO	A service unit responsible for certain aspects of project practice within an organisation.	Widely different charters can be established for PMOs using different sub-sets of a common collection of Terms of Reference. Some of those commonly used Terms of Reference are inconsistent not only with sound practice, but also the theoretical principles presented here.	
Project manager PM	The person held accountable by the project owner(s) for the delivery of the project's outputs and for meeting the project's constraints.	A project can have only one overarching project manager. The project manager is the project owner's supplier. A PM can be contracted (from outside the sponsoring entity).	As project manager, Paul Myer is Nancy's supplier of all the outputs that are included in the project's scoping statement.

(continued)

(continued)

Term	Definition / description	Discussion	Examples & illustrations
Project owner PO	The person(s) held accountable by the funder(s) of the project for securing the project's target outcomes. The project owner acts on behalf of the funder(s) throughout the project—seeking to ensure that their interests are being served.	A project can have multiple owners. The one entity can fill the roles of both funder and owner. The project owner is the project manager's "client". A PO must be appointed from within the sponsoring entity. Refer also: client.	Nancy is, effectively, Paul's "client" (to whom he will notionally deliver all the outputs that are included in the project's scoping statement).
Project plan	A baseline document that provides all the information required to: Make a reliable decision about approving a start to work on a project. Track a project's progress.	The project plan becomes the project's reference model—on which the management of the exercise is based. The project-manager-designate is normally commissioned (by the Project Owner) to develop the project plan, but the project plan belongs to the Project Owner (not the project manager).	
Project supplier	A "contracted" entity who provides component-based inputs or non-salaried labour to the project manager.	Suppliers can be internal or external.	

(continued)

(continued)

Term	Definition / description	Discussion	Examples & illustrations
Project's worth	An overall index of a project's value. The "equation of worth" of a project involves three variables: Benefits. Disbenefits. Costs.	Worth represents a judgement by the funder about the net value of the project. Refer also <u>assessment</u>.	
Quality	A property of an output defined by its list of fitness-for-purpose features.		
Rationale	A statement explaining why this particular <u>project</u> is being proposed at this particular time.		Charles is convinced that if this project is not undertaken immediately, the company's growth ambitions will be seriously handicapped.
Realise (a target outcome)	To successfully generate a target outcome	Outcomes are realised—while outputs are delivered	
Reference group	A reference group is a formally constituted forum of key stakeholders—charged with filling a defined role in the project.	Reference groups report into the project (in the organisational sense) through either the PM or the Steering Committee. Reference groups are distinct from Steering Committees. Reference groups have highly specific Terms of Reference—and so may have a very short tenure.	

(continued)

(continued)

Term	Definition / description	Discussion	Examples & illustrations
Register	A tabular catalogue of a particular class of project entity in which instances are associated with rows and their attributes are shown in columns.	The most common registers in a business case or project plan are: Stakeholder. Issues. Risk. The rows of a register are associated with instances, while the columns are associated with attributes.	
Residual risk	The risk exposure that remains after selected risk mitigation programmes have been adopted.	Also called "residual RE".	
Resolving (an infeasibility)	Taking actions to reduce the level of infeasibility to zero. The available actions depend on both the type and severity of the infeasibility.	All infeasibilities must be resolved before a project can be approved (or progressed). Refer also time-infeasible and cost-infeasible projects.	
Resources	See inputs.		
Resource plan	A component of a project plan—indicating the full scope of resources that will be demanded by the project's Work Breakdown Structure.	There are two parts to a resource plan: An acquisition plan. An HR plan The cost of a project is equal to value of the resource plan. If the project budget is less than the cost of the project, it becomes infeasible.	

(continued)

(continued)

Term	Definition / description	Discussion	Examples & illustrations
Risk	Common use involves two separate but related meanings for "risk": "A risk": A scenario in which a project suffers a damaging impact. "Risk": An overall attribute of a project indicating the reliability of the business case (or project plan).	A risk is a single realisation of an event-impact model. This means that a risk has three components: A threat (triggering event). A chain of consequences. A final (damaging) impact. A risk Refer also: threat, risk exposure.	That Lindsay Thomas retires early is a risk to Project BuyRite. Project BuyRite is regarded by key stakeholders as a medium risk project.
Risk (attribute)	An attribute of a project indicating the level of uncertainty that the values set for parameters in a business case (or project plan) will be realised.	Risk (also called "riskiness") is usually gauged using a qualitative measure such as Risk Exposure (RE). Refer also: risk (noun), threat, risk exposure, riskiness	
Risk (scenario)	A scenario in which the worth of a project falls relative to that set in the currently-approved business case.	Throughout this text we use an event-impact model of a risk. Refer also: risk (attribute), threat, risk exposure.	

(continued)

(continued)

Term	Definition / description	Discussion	Examples & illustrations
Risk Exposure	A qualitative measure of the "importance" of a threat.	RE is similar to—but distinct from the statistical concept of "expected value". Instead of multiplying probability by a damage bill, we use a Grading Table to convert the values we have assigned for Li & Se to RE. RE is expressed in terms of a value from a "wordscale" such as "A", "B", "C", … . The RE of a threat before a Risk mitigation programme RMP is put into place is called "Pre-RE". The RE of a threat after an RMP is put into place is called "Post-RE".	
Risk management	The broad meaning: An analytical/creative procedure for identifying, analysing, tracking and managing risks.	The narrow meaning: The work of managing a risk.	
Risk manager	A member of the project team assigned responsibility for administering the risk management processes surrounding the project.		
Risk Mitigation Programme RMP	A set of actions that will cause the overall RE of the project to fall.	The actions that make up an RMP are of two kinds: Preemptives—that cause Li to fall. Contingencies—that cause Se to fall.	

(continued)

(continued)

Term	Definition / description	Discussion	Examples & illustrations
		An RMP requires additional resources and so has the effect of lowering overall project RE—but with additional cost.	
Risk Register	A tool for cataloguing risks and documenting their analysis and management.	Three participants in a project can enter threats into a risk register: the Project owner, the Project manager, the Project assurance counsellor. Under certain circumstances they may each maintain separate registers.	
Riskiness	A descriptive term to indicate the level of uncertainty that the value set for project's worth in a business case (or project plan) will be realised.	Refer also: risk.	
Risk Report	A regular report that shows "important" risks for consideration by a project Steering Committee.	The project manager is responsible for developing the risk report.	
Role description	A specification of the role to be played in a project by an internal team member.	Refer also: charter, terms of reference.	

(continued)

(continued)

Term	Definition / description	Discussion	Examples & illustrations
(An assessment) rule	Shows how to use measurements taken in an assessment test to judge success or failure.	Refer also: assessment test.	
Scope	A definitive statement about a project's boundaries.	Principle: A project is scoped if and only if its outputs are defined. A project's scope can be determined only if its target outcomes have been established. Refer also setting project scope.	
Scoping statement	A scoping statement for a project has three elements: A statement of objective. A list of target outcomes. A list of outputs with critical fitness-for-purpose characteristics identified for each.		
Schedule	An array of data set out as a table.	Some examples of common project schedules include: Schedule of milestones. Delivery schedule.	
Setting (project) scope	The process of developing a scoping statement for a project.	Refer also: defining (project) outputs.	

(continued)

(continued)

Term	Definition / description	Discussion	Examples & illustrations
Severity Se	A qualitative measure of the level of damage that the project will suffer if a threat occurs.	Severity is gauged using a word from a pre-fabricated wordscale (such as: Negligible, Moderate, Grave). The Se of a threat before a Risk mitigation programme (RMP) is put into place is called "Pre-Se". The Se of a threat after an RMP is put into place is called "Post-Se".	
Spiral (approach to project initiation)	An iterative approach to Initiation and planning in which the business case and project plan are developed iteratively.	The approach is based loosely on an iterative technique proposed by Boehm in the 1970s for software development.	
Sponsor	See <u>Sponsoring entity</u>.		
Sponsoring entity	The organisational unit(s) appointed by the funder to assemble the project's <u>baseline documents</u>—and/or from which the owner will be appointed to oversee the <u>project</u>.	Different sponsoring entities can be appointed for <u>initiation</u> and <u>execution/outcome realisation</u>.	
Spontaneous stakeholding	A form of stakeholding arising when an entity's interest in a	There are six generic classes of spontaneous stakeholder in a project: <u>Beneficiaries</u>.	

(continued)

(continued)

Term	Definition / description	Discussion	Examples & illustrations
	project arises regardless of any appointment to fill a role in the governance model.	Customers. Funders. Impactee (negative). Impactee (positive). Influencers. Refer also: Commissioned stakeholding.	All existing procurement staff at ICO are spontaneous stakeholders in Project BuyRite (because, amongst other things, they are impactees).
Stage	One of a series of sequentially-related sub-projects.	The rationale for the overall project is to be found in the final stage. Each stage has its own outcomes and work. Approval of a stage depends on the achievement of the target outcomes of the preceding stage.	For example: a feasibility study is an early stage of a typical development project. It will produce a report—on which a decision to proceed with later stages will be based.
Stakeholder	An individual or entity who is either: potentially impacted by the project; or, who has a potential impact on the project.	Anyone with an interest in the project is, by definition, a stakeholder in it. The set of all stakeholders is broken up into an exhaustive collection of generic classes—related to the nature of their interest. A stakeholder can be a member of more than one class. Refer also stakeholding	
Stakeholder engagement plan	A collection of activities that will be under-taken in order to	There are three generic forms of stakeholder engagement: Include in the project communication plan.	To engage Lindsay Thomas, Paul Myer has recommended that he: Be included in a regular executive project briefing.

(continued)

(continued)

Term	Definition / description	Discussion	Examples & illustrations
	appropriately engage a project's stakeholder.	Make the subject of a special programme of engagement. Include in the Project Governance Model.	Be appointed into a senior role in the new Procurement Unit when the project is finished. Be included in the PGM as a technical adviser
Stakeholding	The stakeholding of a stakeholder in a project is defined by a list of the two-way impacts between the stakeholder and project.	Refer also stakeholder.	The stakeholding of existing ICO Procurement staff in Project BuyRite includes: They will have their jobs significantly redefined. They are critical to the development of analytical models of the As-Is procurement processes.
Statement of objective	A formal statement about the purpose of the project.	This statement has a number of properties: It is short—so that it can be used as a slogan by project stakeholders. It begins with the word "To …" It is expressed in outcome terms—possibly qualified with a reference to the primary output. It is high-level and general in style. A statement of objective is supported with a set of defined target outcomes. Refer also: target outcomes.	
Steel tetrahedron	A suggested robust alternative (to the "iron triangle") test of success for project management.	The "steel tetrahedron" requires that all the project's outputs be delivered: To an agreed specification. On time. Within budget.	

(continued)

(continued)

Term	Definition / description	Discussion	Examples & illustrations
		Without detrimental outcomes attributable to the actions of the project manager. The steel tetrahedron allows for trade-offs amongst the three criteria and thus establishes necessary and sufficient conditions for project management success. Refer also: "iron" triangle.	
Steering committee SC	The body that is formally charged with supporting the project owner in discharging his/her accountabilities. The SC is made up of a small group of powerful supporters of the project.	The SC's charter is based on a single objective—that of ensuring the project is always "pointed at" its business case. The funder will usually treat the SC as collectively responsible for the project's success. An SC is to be distinguished from a Reference Group. Only the SC can approve changes to the project plan—especially the scoping statement. Anyone opposed to the project is disqualified from SC membership because of the resulting conflict of interest. Refer also Reference group.	
Success	A state of acceptable performance.	Success is judged (or determined). By way of contrast, performance is measured (or gauged).	
Successor	An element of a Work Breakdown Structure that is constrained in time by the execution of another element.	Refer also: dependency; predecessor.	
Supplier	See project supplier.		
Tangibility	A characteristic of an entity whereby it can be touched.	Outputs, as artefacts, are always tangible. Outcomes as end-effects, are never tangible (but they are measurable).	

(continued)

(continued)

Term	Definition / description	Discussion	Examples & illustrations
Target outcomes	A desirable outcome that is sought by the funder—and for which a threshold of achievement is set.	A target outcome is defined when values have been set for the following parameters: Name. Description. Measure. Method of measurement or source of data. Target level. Target date for realisation. Owner (who will be held accountable for realisation).	In project BuyRite a target outcome is "Reduced procurement costs".
Task	A primitive process representing the lowest level of work recognised in a Work Breakdown Structure. Also used loosely to mean any "block of work".	In the stylised three-level Work Breakdown Structure, processes at the third (lowest) level are called tasks. A collection of related tasks constitutes an activity. If a task is assigned to a team member it becomes a work package. Word structure: Tasks are expressed as imperatives (command phrases). Refer also: phase, activity.	
Team member	An entity that provides labour or produces outputs for the project.	Can take either of two forms: An individual or entity who is assigned to the team to provide labour-based inputs to the project. An internal or external entity that is "subcontracted" to produce a specific project output. Team members can be internal or external (to the sponsoring entity).	

(continued)

Term	Definition / description	Discussion	Examples & illustrations
(continued)			
Template	A tool involving a structure of headings or prompts that is used to guide the assembly of a form, register or document.		
Terms of reference ToRs	A specification of the role to be played in a project by an element of the Project Governance Model—other than internal team members.	Refer also: charter, role description.	
The line	A notional boundary that separates the work (and hence the outputs) of a project into two disjoint subsets: above-the-line and below-the-line.	The line forces a dichotomy onto the set of all tasks and outputs associated with a project. Tasks and outputs are either above-the-line or below-the-line—but they cannot be both. This concept of the line in project management is not to be confused with a similar term used in marketing to classify advertising media.	
Threat	An event that has two characteristics: It may or may not happen. If it does happen, it	A triggering event has a property called likelihood (Li). Li is analogous to probability. Threats are further classified as: Realised—if they happen (after the project has started).	"Project manager resigns" is an example of a threat.

(continued)

(continued)

Term	Definition / description	Discussion	Examples & illustrations
	will eventually lead to a decrease in the worth of a project. Also called a "triggering event".	Potential—if they have not yet happened. In this text we do not call a realised threat an "issue". Word structure: Threats are expressed in "newspaper headline" style Refer also: issue.	
Time-infeasible (project)	A project for which an optimal timeframe implied by its workplan exceeds the timeframe agreed-to by the owner/funder.	In such a situation, deadlines fall earlier than the milestones derived from an optimal workplan. The difference between a deadline and the corresponding milestone is a gauge of the level of infeasibility of the project.	
Timeframe	The overall duration of a project.	There are two views of a project's duration: The project manager's—measured from approval through to delivery of all outputs. The funder's/owner's—measured from approval through to realisation of target outcomes.	
Transaction(s)	An execution of a business process.		
Underscoping	A situation in which the current scope of a project is missing critical outputs (or some outputs are missing necessary fitness-for-purpose features).	If a project is underscoped, it cannot realise its target outcomes. Refer also: overscoping.	

(continued)

(continued)

Term	Definition / description	Discussion	Examples & illustrations
Undesirable outcomes	An outcome that reduces value in the eyes of the project funder.	Whereas a desirable outcome can only emerge from a project after its outputs have been delivered, an undesirable outcome can emerge from a project before outputs have been produced. Refer also: outcome.	The dust, noise and short-term congestion arising from the construction of a new bridge are examples of undesirable outcomes associated with that project.
User	A term from IS/IT that is not used in the ITO model.	A corresponding defined term in the ITO model is project customer.	
Utilisation	The employment of an output by a project's customer(s) that can cause target outcomes to emerge as a by-product.		
Utilisation map	A tool used to analyse the robustness, completeness and appropriateness of a project's scoping statement.	A utilisation map is an aggregation of all the project's Utilisation matrices. It takes the form of a table in which: The rows are associated with outputs. The columns are associated with target outcomes. Cell entries take the form of a list of all those project customers who utilise the outputs linked to this row to generate the outcomes linked to this column.	
Value	See flow of value.		
Work Breakdown Structure WBS	A hierarchical model (structured breakdown) of the	A stylised three-level WBS is used for discussion purposes. The layers of this structure are called phase, activity and task	This is an appropriately-worded WBS element (because it is stated as an imperative (command):

(continued)

(continued)

Term	Definition / description	Discussion	Examples & illustrations
	work involved in executing a project. The term "WBS" can be applied to the work involved in producing a single project output—or to the project as a whole.	respectively. In practice WBSs can have any number of levels. In general, all elements of WBSs are simply called "tasks". Refer also: WBS, phase and task.	"Arrange meetings with major existing suppliers" These are unacceptable forms of wording for WBS elements (because they are nouns): "Minutes of meeting with supplier" "Results of process evaluation."
Word structure	The permissible ways in which particular instances of terms from this glossary are to be expressed.	(This is a meta-term within this glossary—it is used to construct the glossary—but is not, itself, a project management term). This requirement ensures that specific instances of terms are used consistently and can be classified unambiguously.	Examples of word structures: Outputs are expressed as nouns. Outcomes are expressed as changes in a measurable variable.
Work package	An element of the Work Breakdown Structure that can be assigned to a team member and later tracked as a single "block of work".	A task is the work package in the stylised, three-level WBS used here.	
Workplan	A comprehensive schedule of the project's work	These are usually based on a diagram showing: The relationships amongst all of the elements of the Work Breakdown Structure. Critical attributes of those elements (such as timings).	The two most common forms of workplan are: The Gantt chart. A network diagram. A workplan is a core component of a project plan.

(continued)

(continued)

Term	Definition / description	Discussion	Examples & illustrations
Workplanning	The process of assigning the elements (phases, activities & tasks) of a WBS to a calendar.	The primary output of workplanning is a workplan.	
Worth	See project worth.		

Appendix B
The Input-Transform-Outcome (ITO) Study

As part of the development of the ITO model (described in Chap. 2), we conducted a study to learn about the perceptions of projects amongst funders and project managers. The objective was to understand the importance attached to different elements by these two key players

A questionnaire was then developed and distributed to 102 managers in Asia Pacific countries asking them to rank 16 factors (described in Table B.1) that had

Table B.1 The study's questionnaire: importance of project management factors to the project's funder

Project management factor	Not important			Extremely important	
Developing a business case	1	2	3	4	5
Approving a business case	1	2	3	4	5
Developing a list of agreed outputs (deliverables)	1	2	3	4	5
Developing a list of agreed target outcomes (benefits)	1	2	3	4	5
Developing a project plan	1	2	3	4	5
Monitoring and controlling the project	1	2	3	4	5
Updating the project plan	1	2	3	4	5
Managing project risks	1	2	3	4	5
Assembling a suitable project team	1	2	3	4	5
Managing the project team	1	2	3	4	5
Developing the project team	1	2	3	4	5
Effective communications with stakeholders	1	2	3	4	5
Support provided by senior managers	1	2	3	4	5
Producing outputs (deliverables)	1	2	3	4	5
Achieving target outcomes (benefits)	1	2	3	4	5
Assigning a person accountable for target outcomes (benefits) achievement	1	2	3	4	5

Table B.2 The ranking of project management factors as are important to funders

Project management factors	Importance to funder
1. Achieving target outcomes (benefits)	3.98
2. Approving a business case	3.92
3. Developing a business case	3.91
4. Developing a list of agreed outputs (deliverables)	3.91
5. Producing outputs (deliverables)	3.91
6. Developing a list of agreed target outcomes (benefits)	3.84
7. Effective communications with stakeholders	3.70
8. Monitoring and controlling the project	3.68
9. Developing a project plan	3.61
10. Managing project risks	3.53
11. Assigning a person accountable for target outcomes (benefits) achievement	3.46
12. Support provided by senior managers	3.40
13. Assembling a suitable project team	3.19
14. Updating the project plan	3.17
15. Managing the project team	3.09
16. Developing the project team	2.70

Table B.3 Effort invested by project managers

Project management factors	Effort invested by project managers
1. Producing outputs (deliverables)	3.93
2. Monitoring and controlling the project	3.81
3. Effective communications with stakeholders	3.80
4. Developing a project plan	3.71
5. Achieving target outcomes (benefits)	3.68
6. Developing a business case	3.64
7. Developing a list of agreed outputs (deliverables)	3.63
8. Managing project risks	3.49
9. Developing a list of agreed target outcomes (benefits)	3.47
10. Approving a business case	3.41
11. Managing the project team	3.37
12. Assigning a person accountable for target outcomes (benefits) achievement	3.36
13. Assembling a suitable project team	3.34
14. Updating the project plan	3.26
15. Support provided by senior managers	3.17
16. Developing the project team	2.87

been identified in the general project management literature (e.g. Kerzner, 2009; PMI, 2008) and also from the framework proposed here. Using a five point Likert scale, managers were asked to: (1) evaluate the importance of these factors to the project funder (see Table B.2) and (2) assess the level of effort invested in each during the project (see Table B.3). Respondents were also asked to provide information about various organisational and project characteristics. 29 responses were received from India, 16 from New Zealand, 15 from Australia and the rest from different Asia Pacific countries. 20.2% of responses came from software organisations, 19.0% from services, 11.9% from engineering, 13.1% from government and 8.3% from production organisations. Project duration ranged between 2 and 60 months with an average of 15.6 months. 55.4% of the projects were internally funded (within the same organisation), while 44.6% of the projects had external funders. 89% of responses were male and 11% female.

Table B.2 presents the mean scores obtained for each factor, on a scale ranging from 1 (not important) to 5 (extremely important).

Table B.2 shows that *achieving target outcomes* is the most important factor for project funders, scoring a mean of 3.98. Additional statistical analysis reveals a 95% confidence interval of (3.78, 4.18) confirming that *achieving target outcomes* is significantly more important to a funder than most other factors.

In the same study, project managers were asked to report on the level of effort dedicated to each of the same 16 project management factors on a scale of 1 (low) to 5 (high). Those results are reported in Table B.3.

These results support the claim that *achieving target outcomes* is the most important factor to a project funder and that project managers invest most of their time in *producing project outputs*.

Appendix C
The Critical Success Processes Study

C.1 The Critical Success Processes (CSP) Study

Executives should ensure not only that Critical Success Processes (CSPs) receive particular attention from project managers, but that they themselves also support these processes appropriately.

Top management support and project planning were selected for this analysis because both are frequently identified as project management Critical Success Factors (CSFs) (Johnson, Karen, Boucher, & Robinson, 2001; Pinto & Slevin, 1988; Zwikael & Globerson, 2006).

The research model and a questionnaire used in this study are presented in Fig. C.1. The model relates the contribution of 33 project related processes to four project success measures (see below) in different project contexts. Industry, country and project focus served as moderating variables in the relationship between the selected processes and level of project success. In the questionnaire, project managers were asked to estimate the frequency with which these processes were applied to recent completed projects (using a 1–5 Likert scale). An approach based on the Learning Curve theory (Griffith, 1996; Snead & Harrell, 1994; Yiming & Hao, 2000; Watson and Behnke, 1991) and the Expectancy Theory Model (Vroom, 1964) was used to evaluate the extent to which each was used. Respondents were asked to indicate the frequency with which the outputs that are produced during execution of each process had been created. To help project managers make reliable estimates, a full explanation of each process was provided in the questionnaire.

While the independent variables were collected from project managers, the dependent variables (related to measures of project success) were obtained from their supervisors to avoid a 'same source bias'. In this study, project success results were reported using the following four indices, (also described in Chap. 3):

1. **Schedule overrun.** The actual project schedule as a percentage of the original plan. This measure represents the "Timeframe" dimension of project management success.

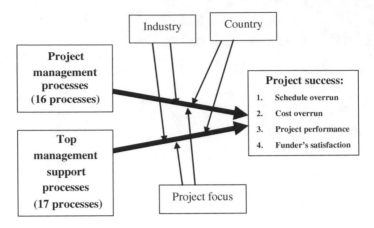

Fig. C.1 The critical success processes (CSPs) research model

2. **Cost overrun.** The actual project cost as a percentage of the original plan. This measure represents the "Cost" dimension of project management success.
3. **Scope/quality.** This is measured on a scale of zero (no outputs have been achieved) to ten (all committed outputs have been delivered fit-for-purpose and to the agreed quality standards). This measure represents the extent to which the project outputs were delivered in accordance with the original agreed quality specification.
4. **The funder's satisfaction with project's outcomes.** While the previous three measures reflect (short-term) project management success, funder satisfaction reflects the long term strategic contribution of the project to the organisation, as perceived by the funder. Such a variable is not unlike others that have been suggested—such as business impact on the organisation and opening new opportunities for the future (Shenhar & Dvir, 2007). In the terms introduced in Chap. 3, this variable is a surrogate for project outcomes, as defined earlier. Funder satisfaction is gauged qualitatively on a scale of zero (low satisfaction) to ten (high satisfaction).

The sample for this study took the form of more than 1,000 projects— (a project being the unit of analysis). Questionnaires were distributed to project managers and their supervisors from different industries in Japan, Israel and New Zealand during the years 2002–2007. These countries were selected to represent different cultures (Asian and Western), sizes of country (population of 4 through 130 million) and geographical locations. Some additional data was collected in China, India and Australia.

To deal with the problem of missing data, responses were ignored if less than 80% of the questions were answered. The resulting 776

Table C.1 Data used in the study

Industry	Israel	Japan	New Zealand	Others	Total
Engineering	44	1	45	8	98
Software	95	78	44	20	237
Production	15	33	15	1	64
Construction	5	0	15	3	23
Communication	37	1	59	2	99
Services	10	10	31	10	61
Government	69	2	91	6	168
Others	0	0	15	11	26
Total	275	125	315	61	776

Table C.2 The critical success processes (CSPs) model validity

Success Measure	R^2	F	p-value
Schedule overrun	0.20	5.02	< 0.001
Cost overrun	0.13	4.08	< 0.001
Project performance	0.32	9.28	< 0.001
Funder's satisfaction	0.29	7.91	< 0.001

useable responses are shown grouped by countries and industries in Table C.1.

The model's reliability was calculated using a number of statistical tests, such as Cronbach alpha. The result (0.93) was considerably higher than the minimum value required by the statistical literature (Hair, 2006). The model's validity was evaluated using four linear regression runs with all project management processes serving as the independent variables against each of the four project success measures (as the dependent variable). These results are presented in Table C.2.

All results are statistically significant with p-values under 0.001— confirming the model's validity. Further discussion of tests for validity and reliability can be found in Zwikael and Globerson (2004). These results suggest that the project management processes included in the model do contribute significantly to each of the four project success measures.

The next section describes the research questionnaire. This is followed by a discussion of results for the two model groups: Section C.3 for top management support and C.4 for planning.

C.2 The Critical Success Processes (CSP) Questionnaire

Please indicate the most suitable answer for each planning product as it relates to the projects you are currently involved in, according to the following scale:

---

5 The product has always been obtained.

4 The product has frequently been obtained.

3 The product has normally been obtained.

2 The product has seldom been obtained.

1 The product has hardly ever been obtained.

A The product has been irrelevant to the projects
     I am currently involved in.

B I do not know whether the product has been obtained.

Part A: Project Planning Processes

Planning product	Never				Always	Do not know	Irrelevant
Project plan	1	2	3	4	5	A	B
Project deliverables	1	2	3	4	5	A	B
WBS (Work Breakdown Structure) chart	1	2	3	4	5	A	B
Project activities	1	2	3	4	5	A	B
PERT or Gantt chart	1	2	3	4	5	A	B
Activity duration estimate	1	2	3	4	5	A	B
Activity start and end dates	1	2	3	4	5	A	B
Activity required resources	1	2	3	4	5	A	B
Resource cost	1	2	3	4	5	A	B
Time-phased budget	1	2	3	4	5	A	B
Quality management plan	1	2	3	4	5	A	B
Role and responsibility assignments	1	2	3	4	5	A	B
Project staff assignments	1	2	3	4	5	A	B
Communications management plan	1	2	3	4	5	A	B
Risk management plan	1	2	3	4	5	A	B
Procurement management plan	1	2	3	4	5	A	B

Part B: Top Management Support Processes

Top management support product	Never				Always	Do not know	Irrelevant
Project-based organisation	1	2	3	4	5	A	B
Extent of existence of project's procedures	1	2	3	4	5	A	B
Appropriate project manager assignment	1	2	3	4	5	A	B
Extent of project procedures update	1	2	3	4	5	A	B
Extent of involvement of the project manager during initiation stage	1	2	3	4	5	A	B
Extent of communication between the project manager and the organisation during the planning phase	1	2	3	4	5	A	B
Extent of existence of project success measurement	1	2	3	4	5	A	B
Extent of supportive project organisational structure	1	2	3	4	5	A	B
Extent of existence of interactive inter-departmental project planning groups	1	2	3	4	5	A	B
Extent of organisational projects resource planning	1	2	3	4	5	A	B
Extent of organisational projects risk management	1	2	3	4	5	A	B
Extent of organisational projects quality management	1	2	3	4	5	A	B
Extent of on-going project management training programmes	1	2	3	4	5	A	B
Extent of project management office involvement	1	2	3	4	5	A	B
Extent of use of standard project management software (e.g. Ms-Project)	1	2	3	4	5	A	B
Extent of use of organisational projects data warehouse	1	2	3	4	5	A	B
Extent of use of new project tools and techniques	1	2	3	4	5	A	B

C.3 Critical Top Management Support Processes

As a form of executive leadership, top management support is considered critical for project success (Fortune & White, 2006). The reason for this may be that organisational issues have a greater impact than technical issues (Luna-Reyes, Zhang, Gil-García, & Cresswell, 2005). As a result, senior managers should be encouraged to lead the way by establishing strategies, setting expectations and supporting project managers.

C.3.1 Top Management Support Processes

Zwikael (in Zwikael & Globerson, 2006; 2008) undertook an investigation into common project management processes and their adoption across industry groups as well as countries. Seventeen top management support processes were identified. For each the main product and a process description are introduced in Table C.3.

Table C.3 Top management support processes

Top management support process	Top management support product	Product description
Development of a project oriented organisation	Project-based organisation	The organisation manages its projects in a formal way. For example, the project is a central activity in the organisation and project managers have the authority to also manage project budget and staff.
Development of project procedures	Extent of existence of project procedures	Formal project procedures are written, approved and implemented by all project managers.
Assignment of project manager	Appropriate project manager assigned	A match is made between project manager assignment and project type. For example, a conservative project manager would not be assigned to an R&D project.
Maintenance of project procedures	Up-to-date project procedures	Project procedures are updated frequently.
Involvement of the project manager during initiation stage	Extent of involvement of the project manager during initiation stage	Project manager is involved in the processes previous to the contract signing or the approval of the project. For example, project managers are involved in the decision on project duration and the technology to be implemented.
Communication between the project manager and the organisation	Extent of communication between the project manager and the organisation during the planning phase	Employees from other departments are involved in project management. For example, quality insurance manager or an individual from procurement department.
Measurement of project management success using success criteria	Adoption of project management success measures	Project success measures are determined by the organisation. For example, project completion by 31st July.
Supportive project organisational structure	Extent of supportive project organisational structure	The organisational structure supports project managers' needs. For example, project-oriented organisation, where project managers have appropriate independence and authority.
Interactive inter-departmental project planning groups	Employment of interactive inter-departmental project planning groups	Project management is performed in groups made up of individuals from various departments. For example, a budget group that includes project manager, procurement manager and an economist.
Project resource management	Formality of project resource planning	Organisation integrates the management of all projects resources. For example, identifying critical resources or mobilisation of resources amongst projects.

(continued)

Table C.3 (continued)

Top management support process	Top management support product	Product description
Project risk management	Formality of projects risk management	Organisation is involved in identifying risks and preparing a mitigation plan. For example, budget priority for critical risks.
Project quality management	Formality of project quality management	Quality assurance department helps project manager in quality planning. For example, quality assurance is conducted to the project plan.
On going project management training programmes	Formality of on-going project management training programmes	Organisation train project managers frequently. For example, project methodology or project software courses.
Involvement of project management office	Extent of project management office involvement	A project management office exists in the organisation and helps project managers. For example, makes templates for project documents.
Use of standard project management software packages	Extent of use of standard project management software (e.g. MS-Project)	Organisation purchased and implemented or developed software that is dedicated to project management. For example, MS-Project or Primavera.
Use of projects data warehouse	Extent of use projects data warehouse	Organisation operates a library that includes data from projects ended. For example, actual duration of tasks in projects.
Use of new project tools and techniques	Extent of use of new project tools and techniques	New tools and techniques are being searched in the organisation. For example, evaluation of TOC methodology or a new version of MS-Project to the organisation

C.3.2 Extent of Use of Top Management Support Processes

Table C.4 presents the frequency of use (from 1 = low to 5 = high) of these processes, including the mean and standard deviation.

The sample mean varies between 2.48 for "Use of projects data warehouse" and 3.82 for "Communication between the project manager and the organisation". The most frequently used top management support processes are:

1. Communication between the project manager and the organisation
2. Involvement of the project manager during initiation stage
3. Use of standard project management software packages
4. Development of project procedures

Table C.4 Extent of use of top management support processes

Top management support process	N	Mean	Standard Deviation
Development of a project oriented organisation	772	3.58	1.26
Development of project procedures	775	3.69	1.18
Assignment of project manager	775	3.65	1.08
Up-to-date project procedures	767	3.00	1.12
Involvement of the project manager during initiation stage	775	3.76	1.07
Communication between the project manager and the organisation	774	3.82	1.01
Measurement of project management success using success criteria	775	3.50	1.16
Supportive project organisational structure	775	3.39	1.10
Interactive inter-departmental project planning groups	771	3.13	1.12
Project resource management	765	3.07	1.17
Project risk management	773	3.02	1.21
Project quality management	767	2.97	1.20
On going project management training programmes	766	2.74	1.16
Involvement of project management office	756	2.83	1.46
Use of standard project management software packages	773	3.73	1.38
Use of projects data warehouse	763	2.48	1.29
Use of new project tools and techniques	768	2.53	1.13

These results show that most organisations avoid performing those top management support processes that require funds, for example implementing an organisational projects data warehouse that collects data from previous projects, or introducing new project tools and techniques sponsoring training programmes for project managers and project team members. Most senior managers tend to perform those support processes that are easy to use and low-cost. However, the impact of such processes on project success is unclear. The next section considers the relationship between various top management support processes and project success to identify the most critical processes.

C.3.3 Critical Top Management Support Processes

In order to identify critical top management support processes, a multivariate regression has been conducted with the 17 top management support processes acting as independent variables and the four project success measures as the dependent variables. The results are presented in Table C.5.

According to this table, the most effective top management support processes are:

1. Assignment of project manager
2. Communication between the project manager and the organisation

Table C.5 Critical top management support processes

Effect	F	Significance level
Intercept	87.365	0.000 **
Development of a project oriented organisation	0.629	0.642
Development of project procedures	1.262	0.284
Assignment of project manager	7.970	0.000 **
Up-to-date project procedures	4.969	0.001 **
Involvement of the project manager during initiation stage	1.242	0.292
Communication between the project manager and the organisation	10.989	0.000 **
Measurement of project management success using success criteria	4.451	0.001 **
Supportive project organisational structure	3.242	0.012 *
Interactive inter-departmental project planning groups	5.592	0.000 **
Project resource management	0.653	0.625
Project risk management	1.650	0.160
Project quality management	2.159	0.072
On going project management training programmes	3.108	0.015 *
Involvement of project management office	1.344	0.252
Use of standard project management software packages	4.590	0.001 **
Use of projects data warehouse	0.729	0.573
Use of new project tools and techniques	4.769	0.001 **

*$p < 0.05$; **$p < 0.01$

3. Interactive inter-departmental project planning groups
4. Measurement of project management success using success criteria

C.3.4 Critical Top Management Support Processes across Industries

Project management is practiced in different ways among various industries (Cooke-Davies & Arzymanow, 2002; Ibbs & Kwak, 2000; Pennypacker & Grant, 2003). As a result, those top management support processes that are most beneficial across specific industries are of considerable interest.

A multivariate regression was conducted for each industry separately, with the 17 top management support processes acting as independent variables and the four project success measures as the dependent variables. Table C.6 summarises these results.

C.3.5 Critical Top Management Support Processes across Cultures

Various top management support processes differ in their effectiveness across industries. The same research analysis method (Zwikael, 2008) has also revealed differences across cultures. Table C.7 presents those critical top management

Table C.6 Critical top management support processes across industries

Effect / Industry type:	Engineering	Software	Production	Construction	Communications	Services	Government
Intercept	0.000 **	0.000 **	0.000 **	0.028 *	0.000 **	0.000 **	0.000 **
Development of a project oriented organisation	0.829	0.677	0.691	0.519	0.435	0.074	0.939
Development of project procedures	0.575	0.307	0.368	0.047 *	0.229	0.759	0.033 *
Assignment of project manager	0.023 *	0.004 **	0.140	0.214	0.131	0.703	0.050 *
Refreshment of project procedures	0.794	0.011 *	0.509	0.036 *	0.115	0.754	0.323
Involvement of the project manager during initiation stage	0.263	0.069	0.235	0.152	0.332	0.161	0.465
Communication between the project manager and the organisation	0.669	0.000 **	0.036 *	0.054	0.214	0.095	0.069
Measurement of project management success using success criteria	0.224	0.000 **	0.568	0.217	0.057	0.676	0.108
Supportive project organisational structure	0.157	0.018 *	0.333	0.029 *	0.165	0.582	0.001 **
Interactive inter-departmental project planning groups	0.937	0.003 **	0.280	0.036 *	0.011 *	0.219	0.014 *
Project resource management	0.025 *	0.003 **	0.147	0.468	0.876	0.274	0.024 *
Project risk management	0.259	0.137	0.196	0.171	0.058	0.737	0.300
Project quality management	0.318	0.376	0.097	0.015 *	0.326	0.312	0.728
On going project management training programmes	0.137	0.577	0.345	0.027 *	0.135	0.725	0.028 *
Involvement of project management office	0.377	0.021 *	0.903	0.071	0.084	0.010 **	0.055
Use of standard project management software packages	0.057	0.006 **	1.000	0.262	0.119	0.505	0.438
Use of projects data warehouse	0.530	0.365	0.038 *	0.064	0.134	0.737	0.286
Use of new project tools and techniques	0.606	0.084	0.611	0.566	0.051	0.577	0.921

$* p < 0.05; ** p < 0.01$

Table C.7 Critical top management support processes across countries

Effect	Japan		New Zealand		Israel	
	F	Sig.	F	Sig.	F	Sig.
Intercept	12.554	0.000 **	1.563	0.000 **	64.356	0.000 **
Development of a project oriented organisation	1.681	0.160	0.192	0.073	0.422	0.793
Development of project procedures	1.004	0.409	2.400	0.185	0.955	0.433
Assignment of project manager	2.067	0.090	0.348	0.943	3.060	0.017 *
Up-to-date project procedures	0.671	0.234	6.539	0.051	1.643	0.164
Involvement of the project manager during initiation stage	0.737	0.074	1.630	0.845	0.504	0.733
Communication between the project manager and the organisation	1.804	0.613	0.300	0.000 **	1.457	0.216
Measurement of project management success using success criteria	0.443	0.569	5.050	0.168	2.293	0.060
Supportive project organisational structure	2.124	0.134	2.529	0.878	1.702	0.150
Interactive inter-departmental project planning groups	0.654	0.777	1.592	0.001 **	0.704	0.590
Project resource management	2.682	0.083	2.372	0.042 *	0.361	0.836
Project risk management	3.719	0.625	1.111	0.178	0.358	0.839
Project quality management	1.345	0.036 *	1.466	0.053	2.935	0.021 *
On going project management training programmes	0.751	0.007 **	2.222	0.353	2.294	0.060
Involvement of project management office	0.307	0.258	1.526	0.214	1.700	0.150
Use of standard project management software packages	2.815	0.560	1.426	0.068	1.830	0.123
Use of projects data warehouse	0.307	0.873	1.526	0.196	1.116	0.350
Use of new project tools and techniques	2.815	0.029 *	1.426	0.226	2.690	0.032 *

$*p < 0.05$; $**p < 0.01$

support processes for the three countries represented in the study's sample: Japan, New Zealand and Israel.

Cultural differences in project management among countries can be explained using the GLOBE study (House et al., 2004), which leads the cultural diversity theoretical framework:

- Japan—according to the GLOBE study, Japanese have the highest future orientation score among the three countries. Because Japanese managers tend to make decisions that support them in the long term, it has been found in this study that investing in project management training is a unique critical success top management support process in Japan.

- New Zealand—according to the cultural diversity theory, New Zealanders are a collective (rather than individualist) society. This may explain the results of this study, according to communications and interdepartmental work groups most contribute towards success.
- Israel—according to the GLOBE study, Israelis have the lowest power distance score among the three countries. Because Israeli employees tend to ignore the power distance in the organisation, it has been found in this study that the most effective decision executives can make is to appoint the most appropriate individual to manage a project.

C.3.6 Critical Top Management Support Processes for Achieving Different Project Success Measures

In common methodological practice, separate project success criteria are aggregated into a single, overarching measure success (Scott-Young & Samson, 2008). However, to investigate whether different success factors drive different project outcomes (Cohen, Ledford, & Spreitzer, 1996), an analysis of individuated success measures was undertaken. For example, it is expected that the role of top management support in projects that are under a very strict budget or timeframe will differ from projects that are focused on achieving key strategic goals.

A series of multivariable linear regression models was assembled in which one project success measure at a time was used as the dependent variable. Table C.8 presents the effect of the various top management support processes on each project success measure (with significance levels shown in brackets).

These results show that the effectiveness of various top management support processes depends on the success measure that is considered most important for a project. For example, the processes having the greatest impact on funder satisfaction (related to overall project success) are those related to communication between the project managers and senior managers and those concerned with updating project procedures. However, when project management success is most important (concerned with project efficiency), the most effective top management support derives from a clear definition of those same success measures, involvement of the project management office and the availability of a project management software package.

C.4 Critical Planning Processes

This section analyses the relative importance of 16 planning processes across project scenarios, using the same structure applied for top management support processes in Sect. C.3.

Table C.8 Critical top management support processes across project aims

	Schedule overrun		Cost overrun		Project performance		Funder satisfaction	
R square	0.111		0.051		0.274		0.231	
Adjusted R Square	0.088		0.068		0.256		0.212	
F value	4.946		3.951		15.069		12.004	
Sig level	0.000**		0.000**		0.000**		0.000**	
	Beta	Sig	Beta	Sig	Beta	Sig	Beta	Sig
(Constant)		0.000**		0.000**		0.000**	0.231	0.000**
Development of a project oriented organisation	-0.056	0.207	-0.026	0.559	-0.006	0.882	0.026	0.530
Development of project procedures	0.069	0.150	-0.018	0.712	-0.024	0.580	-0.005	0.905
Assignment of project manager	0.033	0.486	-0.009	0.855	0.230	0.000**	0.132	0.003**
Up-to-date project procedures	-0.006	0.903	0.040	0.403	-0.175	0.000	-0.147	0.001
Involvement of the project manager during initiation stage	-0.009	0.846	-0.033	0.456	-0.007	0.861	0.068	0.098
Communication between the project manager and the organisation	0.062	0.180	0.065	0.166	0.251	0.000**	0.231	0.000**
Measurement of project management success using success criteria	-0.156	0.000**	-0.162	0.000**	0.020	0.609	0.054	0.178
Supportive project organisational structure	-0.105	0.029*	-0.077	0.116	-0.088	0.043	0.003	0.941
Interactive inter-departmental project planning groups	0.069	0.113	0.066	0.136	0.170	0.000**	0.107	0.008**
Project resource management	-0.032	0.504	-0.027	0.570	0.065	0.128	0.045	0.308
Project risk management	-0.107	0.034*	-0.104	0.042*	0.005	0.912	-0.039	0.410
Project quality management	-0.099	0.059	-0.006	0.913	0.067	0.157	0.077	0.110
On going project management training programmes	-0.006	0.902	0.005	0.913	-0.140	0.001	-0.089	0.034
Involvement of project management office	-0.070	0.110	-0.090	0.042*	-0.013	0.733	-0.033	0.415
Use of standard project management software packages	0.093	0.031	0.064	0.147	0.130	0.001**	0.113	0.005**
Use of projects data warehouse	-0.024	0.586	-0.012	0.790	-0.066	0.102	-0.062	0.136
Use of new project tools and techniques	0.064	0.191	-0.018	0.708	0.148	0.001**	0.143	0.002**

*p < 0.05; **p < 0.01

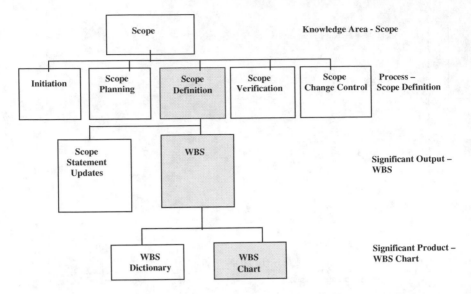

Fig. C.2 Planning processes and products within the scope knowledge area

C.4.1 Project Planning Processes

Planning processes included in this model were identified from the PMBOK (PMI, 2008). Because many of these processes have multiple outputs (products), the most important of these was identified (Zwikael & Globerson, 2004).

Figure C.2 identifies the product to emerge from the planning process. For example, the PMBOK knowledge area called "Scope", includes two planning processes—"Scope Planning" and "Scope Definition". The "Initiation" process is performed before the formal start of project planning, while "scope verification" and "scope change control" are part of the controlling processes.

Figure C.2 uses the planning process called "Scope Definition" (explained below) as an example. A major assumption of the model is that the quality of the output is a function of the frequency in which this output is generated. The justification for this assumption is based on learning theory; "Learning Curve" research has shown that there is an ongoing improvement of performance as a function of the number of times the operation is repeated (e.g. Griffith, 1996; Snead & Harrell, 1994; Watson & Behnke, 1991; Yiming & Hao, 2000). Furthermore, the "Expectancy Theory Model" claims that one will not repeat a process that has no significant added value to one's objectives (Vroom, 1964). Tatikonda and Montoya-Weiss (2001) found that achievement of operational outcomes in 120 development projects aids the achievement of market outcomes. Finally, although much is said today about controlling the processes rather than the outputs (for instance, see the entire ISO9000 series), some control models suggest "output oriented control" when it comes to operational processes, such as project management (Veliyath, Hermanson, & Hermanson, 1997).

In light of the above, an evaluation of the quality of planning processes in this model is based on the frequency of generating the desired outputs and the desired products derived from them. For example, there are two outputs in the "Scope Definition" process: the "WBS" (Work Breakdown Structure) and the "Scope Statement Updates" (see Fig. C.2). The "WBS", deals with the identification of the components of the project's work. The WBS is a new output, which has not been generated from of another process. The same is not true of "Scope Statement Updates", where the output takes the form of an updated entity that has already been generated by another process. Moreover, there are two products included in the "WBS" output—the "WBS Chart", which breaks down the project into manageable work packages and the "WBS Dictionary", which specifies the content of each work package. The "WBS Dictionary" is actually an expansion of the "WBS Chart". Therefore, one may say that the "WBS Chart" is the major product, from which the other is derived. Following that methodology, one major product was defined for each planning process included in the PMBOK.

Table C.9 shows the list of the 16 planning processes and products included in the model.

C.4.2 Extent of Use of Project Planning Processes

Table C.10 presents the frequency of use (from 1 = low to 5 = high) of these processes, including the mean and standard deviation.

The most frequently performed planning processes are:

1. Scope planning
2. Project plan development
3. Schedule development
4. Activity duration estimating

C.4.3 Critical Project Planning Processes

In order to identify critical top management support processes, a multivariate regression has been conducted with the 16 planning processes acting as independent variables and the four project success measures as the dependent variables. The results are presented in Table C.11.

The most effective planning processes are:

1. Staff acquisition
2. Project plan development
3. Cost estimating
4. Activity definition

Table C.9 The 16 planning processes and products included in the model and their related knowledge areas

Knowledge area	Planning process	Planning product	Product description
Integration	Project plan development	Project plan	The document includes all planning products and is used as a management tool to the execution phase. For example, project deliverables, activity start and end dates and role and responsibility assignments.
Scope	Scope planning	Project deliverables	A clear description of the various products that should be achieved when project ends. E.g. a three-module software, documentation and training.
	Scope definition	WBS (work breakdown structure) chart	Hierarchical chart of all activities needed to be performed during execution phase of the project.
Time	Activity definition	Project activities	Description of all activities that should be performed in execution phase of the project. The list includes small and manageable components and their detailed description. For example, acceptance tests.
	Activity sequencing	PERT or Gantt chart	A chart represents project activities and their dependencies. For example, acceptance tests can't start before integration tests end.
	Activity duration estimating	Activity duration estimates	A quantitative estimation of the duration needed to complete the execution of all activities in the project. For example, 10 working days for acceptance tests.
	Schedule development	Activity start and end dates	Definitions of start and end planning dates for each activity in the project. Usually presented in a Gantt chart. For example, acceptance tests between February 1st and February 14th.

(continued)

Table C.9 (continued)

Knowledge area	Planning process	Planning product	Product description
Cost	Resource planning	Activity required resources	Amount of resources and its type required to the execution of each activity in the project. For example, 2 full-time programmers and an external consultant for acceptance tests.
	Cost estimating	Resource cost	Estimation of the cost for each resource in the project. For example, $100 per hour for an external consultant.
	Cost budgeting	Time-phased budget	Presentation of project cost over time. For example, monthly project cost.
Quality	Quality planning	Quality management plan	The document describes the implementation of quality policy in the project, including processes, procedures, responsibility and resources.
Human Resources	Organisational planning	Role and responsibility assignments	Identification of the responsible team member for each project activity. For example, QA manager for acceptance tests.
	Staff acquisition	Project staff assignments	All fitted resources were assigned to the project team.
Communications	Communications planning	Communications management plan	The document describes the formal communication in the project. It includes the methods to be used to gather and store various types of information, distribution lists etc.
Risk	Risk management planning	Risk management plan	The document describes risks that may damage project success, their scoring and a response plan.
Procurement	Procurement planning	Procurement management plan	The document describes the plan for solicitation or contractor management. For example, contract type to use in the project.

Table C.10 Extent of use of planning processes

Planning process	N	Mean	Std. Deviation
Project plan development	776	4.18	1.06
Scope planning	775	4.22	0.96
Scope definition	776	3.70	1.24
Activity definition	773	4.05	1.00
Activity sequencing	774	3.83	1.28
Activity duration estimating	776	4.13	1.02
Schedule development	776	4.15	0.96
Resource planning	773	3.79	1.10
Cost estimating	771	3.60	1.30
Cost budgeting	766	3.39	1.31
Quality planning	776	3.10	1.21
Organisational planning	774	3.82	1.07
Staff acquisition	766	3.67	1.09
Communications planning	773	3.09	1.32
Risk management planning	775	3.33	1.32
Procurement planning	753	2.95	1.21

Table C.11 Critical planning processes

Effect	F	Sig.
Intercept	87.017	0.000 **
Project plan development	7.380	0.000 **
Scope planning	1.042	0.384
Scope definition	1.692	0.150
Activity definition	3.769	0.005 **
Activity sequencing	1.167	0.324
Activity duration estimating	2.345	0.053
Schedule development	2.548	0.038 *
Resource planning	3.094	0.015 *
Cost estimating	6.450	0.000 **
Cost budgeting	1.310	0.265
Quality planning	3.435	0.009 **
Organisational planning	0.591	0.670
Staff acquisition	8.616	0.000 **
Communications planning	2.407	0.048 *
Risk management planning	3.548	0.007 **
Procurement planning	2.895	0.021 *

$*p < 0.05$; $**p < 0.01$

C.4.4 Critical Project Planning Processes across Industries

A multivariate regression model was assembled for each industry, with the 16 planning processes acting as independent variables and the four project success measures as the dependent variables. Table C.12 summarises these results.

Table C.12 Critical planning processes across industries

	Engineering	Software	Production	Construction	Communications	Services	Government
R squared	0.306	0.153	0.507	0.893	0.296	0.568	0.247
Intercept	0.000 **	0.000 **	0.000 **	0.046 *	0.000 **	0.000 **	0.000 **
Project plan development	0.042 *	0.070	0.020 *	0.178	0.156	0.099	0.024 *
Scope planning	0.643	0.109	0.115	0.154	0.667	0.623	0.538
Scope definition	0.574	0.085	0.165	0.093	0.236	0.999	0.830
Activity definition	0.148	0.020 *	0.014 *	0.052	0.523	0.776	0.005 **
Activity sequencing	0.277	0.424	0.389	0.034 *	0.859	0.706	0.057
Activity duration estimating	0.285	0.876	0.117	0.070	0.748	0.390	0.018 **
Schedule development	0.619	0.874	0.733	0.705	0.268	0.036 *	0.076
Resource planning	0.365	0.043 *	0.421	0.309	0.992	0.470	0.023 *
Cost estimating	0.809	0.004 **	0.106	0.524	0.001 **	0.236	0.701
Cost budgeting	0.218	0.425	0.128	0.129	0.562	0.887	0.929
Quality planning	0.929	0.131	0.675	0.232	0.565	0.105	0.559
Organisational planning	0.708	0.208	0.106	0.112	0.103	0.215	0.217
Staff acquisition	0.612	0.191	0.002 **	0.065	0.071	0.384	0.064
Communications planning	0.685	0.250	0.314	0.344	0.124	0.211	0.077
Risk management planning	0.828	0.016 *	0.650	0.013 *	0.213	0.420	0.133
Procurement planning	0.092	0.426	0.068	0.159	0.531	0.746	0.086

*p < 0.05; **p < 0.01

Table C.13 Critical planning processes across countries

	Japan		New Zealand		Israel	
R squared	0.181		0.261		0.129	
Effect	F	Sig.	F	Sig.	F	Sig.
Intercept	15.444	0.000 **	54.977	0.000 **	53.722	0.000 **
Project plan development	1.638	0.170	2.420	0.049 *	0.577	0.680
Scope planning	2.799	0.030 *	0.451	0.772	0.950	0.435
Scope definition	1.639	0.170	0.647	0.629	0.628	0.643
Activity definition	1.763	0.142	1.961	0.101	5.872	0.000 **
Activity sequencing	0.481	0.750	0.923	0.452	1.244	0.293
Activity duration estimating	1.705	0.154	1.714	0.148	0.290	0.885
Schedule development	0.181	0.948	2.194	0.071	2.344	0.055
Resource planning	0.920	0.455	0.827	0.509	1.335	0.257
Cost estimating	1.715	0.152	0.738	0.567	0.404	0.806
Cost budgeting	0.618	0.651	1.038	0.388	1.181	0.319
Quality planning	3.007	0.022 *	4.385	0.002 **	0.401	0.808
Organisational planning	0.090	0.985	0.501	0.735	1.073	0.371
Staff acquisition	4.937	0.001 **	1.949	0.103	0.886	0.473
Communications planning	0.136	0.969	0.936	0.444	0.128	0.972
Risk management planning	2.114	0.084	1.593	0.177	1.410	0.231
Procurement planning	0.934	0.447	3.747	0.006 **	1.762	0.137

$*p < 0.05; **p < 0.01$

C.4.5 Critical Project Planning Processes across Cultures

This section introduces differences across cultures using the same research analysis method used in the previous sections. Table C.13 summarises these results.

C.4.6 Critical Planning Processes for Achieving Different Project Success Measures

This section investigates whether for different project focus (that is dissimilar dominant project success measure) we can identify different planning processes on which managers should focus. A multivariable linear regression model was used with one project success measure serves at a time as the dependent variable. Table C.14 ranks the various planning processes for each project success measure (with significance levels in brackets).

Table C.14 Critical planning processes across project aims

	Schedule overrun		Cost overrun		Project performance		Funder satisfaction	
R squared	0.144		0.137		0.177		0.165	
Adjusted R Square	0.124		0.117		0.158		0.146	
F value	7.229		6.764		9.252		8.536	
Sig level	0.000 **		0.000 **		0.000 **		0.000 **	
	Beta	Sig	Beta	Sig	Beta	Sig	Beta	Sig
(Constant)		0.000 **		0.000 **		0.000 **		0.000 **
Project plan development	-0.033	0.480	-0.056	0.243	0.212	0.000 **	0.242	0.000 **
Scope planning	0.030	0.529	0.014	0.767	0.010	0.832	-0.065	0.169
Scope definition	0.007	0.875	0.027	0.525	-0.077	0.062	0.000	0.994
Activity definition	-0.018	0.701	-0.097	0.037 *	0.134	0.003 **	0.106	0.020 *
Activity sequencing	-0.064	0.144	-0.039	0.371	-0.069	0.106	-0.039	0.366
Activity duration estimating	0.071	0.157	0.078	0.125	0.068	0.165	0.127	0.010 **
Schedule development	-0.157	0.003 **	-0.140	0.008 **	0.001	0.989	0.018	0.727
Resource planning	0.115	0.016	0.155	0.001	0.028	0.541	-0.026	0.585
Cost estimating	-0.133	0.005 *	-0.158	0.001 **	-0.165	0.000	-0.074	0.116
Cost budgeting	-0.068	0.151	-0.038	0.420	0.051	0.270	0.086	0.064
Quality planning	-0.138	0.004 **	-0.040	0.405	0.071	0.131	0.002	0.973
Organisational planning	-0.033	0.482	-0.048	0.311	-0.046	0.321	-0.017	0.715
Staff acquisition	0.034	0.454	0.092	0.046	0.224	0.000 **	0.224	0.000 **
Communications planning	-0.150	0.003 **	-0.071	0.154	-0.006	0.902	0.001	0.984
Risk management planning	0.011	0.829	-0.108	0.034 *	0.015	0.767	-0.077	0.124
Procurement planning	0.112	0.009	0.002	0.964	-0.001	0.976	-0.013	0.758

*p < 0.05; **p < 0.01

References

Ahn, M., & Meeks, M. (2007). Building a conducive environment for life science based entrepreneurship and industry clusters. *Journal of Commercial Biotechnology, 14*, 20–30.

Andersen, E. S., Grude, K. V., & Haug, T. (1995). *The Goal Directed Project Management* (Second ed.). London: Kogan Page.

Angra, S., Schgal, R., & Noori, Z. S. (2008). Cellular manufacturing—A time-based analysis to the layout problem. *International Journal of Production Economics, 112*(1), 427.

APM—Association for project management. (2000). *Body of knowledge* (4th ed ed.). High Wycombe: Association for project management.

Baker, B. N., Murphy, D. C., & Fisher, D. (1988). Factors affecting project success. In D. I. Cleland & W. R. King (Eds.), *Project Management Handbook* (pp. 902–919). New York: Van Nostrand Reinhold.

Barron, J. (2004). This ship is so big, The Verrazano Cringes. The New York Times. April 18.

Boehm, B. (1988). A Spiral Model of Software Development and Enhancement. *Computer, 21*(5), 61–72.

Bourne, L., & Walker, D. H. T. (2005). Visualising and mapping stakeholder influence. *Management Decision, 43*(5/6), 649–660.

Bourne, L., & Walker, D. H. T. (2004). Advancing project management in learning organizations. *The Learning Organization, 11*(2/3), 226–243.

Campbell, H., & Brown, R. (2003). Benefit—Cost analysis. Financial & economic appraisal using spreadsheets, Cambridge.

Cannon-Bowers, J. A., & Salas, E. (1998). Team performance and training in complex environments: Recent findings from applied research. *Current Directions in Psychological Science, 7*(3), 83–87.

Cicmil, S., Williams, T., Thomas, J., & Hodgson, D. (2006). Rethinking Project Management: Researching the actuality of projects. *International Journal of Project Management, 24*(8), 675.

Cohen, S. G., Ledford, G. E. J., & Spreitzer, G. M. (1996). A predictive model of self-managing work team effectiveness. *Human Relations, 49*(5), 643–676.

Cooke-Davies, T. J., & Arzymanow, A. (2002). The maturity of project management in different industries: an investigation into variations between project management models. *International Journal of Project Management, 21*, 471–478.

Crawford, L., Pollack, J., & England, D. (2006). Uncovering the trends in project management: Journal emphases over the last 10 years. *International Journal of Project Management, 24*(2), 175–184.

Cui, Y., & Olsson, N. (2009). Project flexibility in practice: An empirical study of reduction lists in large governmental projects. *International Journal of Project Management, 27*(5), 447.

Daft, R. L. (2007). *Understanding the Theory and Design of Organizations*. Thomson: South-Western.

Dai, C. X., & Wells, W. G. (2004). An exploration of project management office features and their relationship to project performance. *International Journal of Project Management, 22*, 523–532.

Daniel, R. H. (1961). Management data crisis. Harvard Business Review. Sept-Oct, 111-112.

Donaldson, L. (2001). *The contingency theory of organizations*. Thousands Oaks, CA: Sage.

Dunar, A. J., & Waring S. P. (1999). Power To Explore—History of Marshall Space Flight Center 1960–1990. U.S. Government Printing Office. ISBN 0-16-058992-4. Chapter 12.

Dvir, D. (2005). Transferring projects to their final users: The effect of planning and preparations for commissioning on project success. *International Journal of Project Management, 23*(4), 257.

Dvir, D., & Lechler, T. (2004). Plans are nothing, changing plans is everything: the impact of changes on project success. *Research Policy, 33*, 1–15.

Dvir, D., Sadeh, A., & Pines, A. M. (2006). The relationship between Project managers' personality, project types and project success. *Project Management Journal, 37*(5), 36–48.

Dvir, D., Lipovetsky, S., Shenhar, A. J., & Tishler, A. (2003). What is really important for project success? A refined, multivariate, comprehensive analysis. *International Journal of Management & Decision Making, 4*(4), 382.

Eisenhardt, K. M. (1989). Agency Theory: An Assessment and Review. *The Academy of Management review, 14*(1), 57–74.

El-Gohary, N. M., Osman, H., & El-Diraby, T. E. (2006). Stakeholder management for public private partnerships. *International Journal of Project Management, 24*(7), 595–604.

Elias, A. A. (2008). Towards a shared systems model of stakeholders in environmental conflict. *International Transactions in Operational Research, 15*(2), 239–253.

Elias, A., & Zwikael, O. (2007a). *Stakeholder participation in project management: A New Zealand study*. Melbourne, Australia: The 5th ANZAM Operations Management Symposium.

Elias, A., & Zwikael, O. (2007b). *Stakeholder participation in project management: A New Zealand study*. Melbourne, Australia: The 5th ANZAM Operations Management Symposium.

Engwall, M. (2003). No project is an island: linking projects to history and context. *Research Policy, 32*(5), 789–808.

Flyvbjerg, B. (2005). Design by deception: The politics of megaproject approval. *Harvard Design Magazine, Spring/Summer, 22*, 50–59.

Flyvbjerg, B. (2007). Policy and planning for large-infrastructure projects: Problems, causes, cures. *Environment and Planning B: Planning and Design, 34*, 578–597.

Ford, J. D., Ford, L. W., & D'Amelio, A. (2008). Resistance to change: the rest of the story. *The Academy of Management Review, 33*(2), 362–377.

Fortune, J., & White, D. (2006). Framing of project critical success factors by a systems model. *International Journal of Project Management, 24*(1), 53–65.

Fraser, I. (2003). Benefits Realisation: Balancing Outputs with Outcomes. Project Insight—Managing Projects to Realise Strategic Initiatives Conference, Sydney, Australia.

Gardiner, P. D., & Stewart, K. (2000). Revisiting the golden triangle of cost, time and quality: The role of NPV in project control, success and failure. *International Journal of Project Management, 18*(4), 251–256.

Gray, C. F., & Larson, E. W. (2006). Project management—the managerial process, third edition, McGraw-Hill.

Griffith, T. L. (1996). Negotiating Successful Technology Implementation–A Motivation Perspective. *Journal of Engineering & Technology Management, 13*(1), 29–53.

Hackman, J. R. (1987). The design of teams. International J. Lorsch, Editor, Handbook of Organizational Behaviour, Prenctice-Hall, Englewood Cliffs, NJ, pp. 315–342.

Hair, J. F. (2006). *Marketing research*. New York: McGraw-Hill.

Hill, G. M. (2008). *The complete project management office handbook* (2nd ed.). Boca Raton, FL: Auerbach Publications, Taylor & Francis Group.

Hobbs, B., & Aubry, M. (2007). A multiphase research program investigating project management offices (PMOs): The results of phase 1. *Project Management Journal, 38*, 74–86.

Hofstede, G. (2001). *Culture's consequences* (2nd ed.). Newbury Park (CA): Sage Publications.

House, R., Javidan, M., Hanges, P., & Dorfman, P. (2002). Understanding cultures and implicit leadership theories across the globe: an introduction to project GLOBE. *Journal of World Business, 37*(1), 3–10.

Ibbs, C. W., & Kwak, Y. H. (2000). Assessing project management maturity. *Project Management Journal, 31*(1), 32–43.

IPMA. (2006). *International Project management Association competence Baseline, version 3*. Nijkerk, the Netherlands: International Project management Association.

Jassawalla, A. R., & Sashittal, H. C. (1999). Building collaborative cross-functional new product teams. *Academy of Management Executive, 13*(3), 50–63.

Jha, K. N., & Iyer, K. C. (2007). Commitment, coordination, competence and the iron triangle. *International Journal of Project Management, 25*(5), 527–540.

Johnson, J., Karen, D., Boucher, K. C., & Robinson, J. (2001). The criteria for success. *Software Magazine, 21*(1), S3–S11.

Jorgensen, M., & Molokken-Ostvold, K. (2006). How large are software cost overruns? A review of the 1994 CHAOS report. *Information and Software System, 48*(4), 297–301.

Jugdev, K. (2004). Through the looking glass: examining theory development in project management with the resource-based view lens. *Project Management Journal, 35*(3), 15–26.

Kaplan, R. S., & Norton, D. P. (2005). The Balanced Scorecard: measures that drive performance. *Harvard Business Review, 83*(7), 172.

Keller, R. T. (2001). Cross-functional project groups in research and new product development: Diversity, communications, job stress, and outcomes. *Academy of Management Journal, 44*, 547–555.

Kellogg Foundation. (2004). *Logic Model Development Guide*. Michigan: W.K. Kellogg Foundation.

Kerzner, H. (2009). Project Management: A Systems Approach to Planning, Scheduling and Controlling. 10th edition, John Wiley and Sons.

Krooshof, R., Swinkels, F., & Van der Wal, B. (1992). Guide to project management—Prodosta. English translation by Rijkeboer, A. and Thackwray, J. D., Philips, Eindhoven.

Latham, G. P., Erez, M., & Locke, E. (1988). Resolving scientific disputes by the joint design of crucial experiments by the antagonists: Application to the Erez-Latham dispute regarding participation in goal setting. *Journal of Applied Psychology, 73*, 753–772.

Lechler, T., & Dvir, D. (2010). An Alternative Taxonomy of Project Management Structures: Linking Project Management Structures and Project Success. *IEEE Transactions on Engineering Management, 57*(2), 198–210.

Letavec, C. J. (2006). *The program management office: establishing, managing and growing the value of a PMO*. Fort Lauderdale, FL: J. Ross Publishing.

Lim, C. S., & Mohamed, M. Z. (1999). Criteria of project success: an explanatory re-examination. *International Journal of Project Management, 17*(4), 243–248.

Lipke, W. H. (2009). Earned Schedule. Lulu Publishing.

Lipovetsky, S., Tishler, A., Dvir, D., & Shenhar, A. (1997). The relative importance of project success dimensions. *R & D Management, 27*(2), 97–106.

Locke, E., & Latham, G. P. (1990). *A theory of goal setting and task performance*. Englewood Cliffs, NJ: Prentice Hall.

Luna-Reyes, L. F., Zhang, J., Gil-García, J. R., & Cresswell, A. M. (2005). Information systems development as emergent socio-technical change: a practice approach. *European Journal of Information Systems, 14*(1), 93–105.

Malach-Pines, A., Dvir, D., & Sadeh, A. (2009). Project manager-project (PM-P) fit and project success. *International Journal of Operations & Production Management, 29*(3), 268.

McKinsey. (2004). What global executives think. *The McKinsey Quarterly, 2*, 14-25.

McManus, J. (2004). *Risk management in software development projects*. Oxford, UK: Elsevier, Butterworth-Heinemann.

Meredith, J. (2002). Developing project management theory for managerial application: the view of a research journal's editor. Paper presented at PMI Frontiers of Project Management and Research Conference. Seattle, Washington.

Meredith, J. R., & Mantel, S. J. (2009). Project Management—A Managerial Approach. (7th Ed.) John Wiley and Sons.

Miller, G. A. (1956). The magical number seven-plus or minus two: some limits on our capacity for processing information. *The Psychological Review, 63*, 2.

Mintzberg, H. (1979). *The structuring of organisations—A synthesis of research*. NJ: Prentice-Hall.

Morris, P. W. G., Jamieson, A., & Shepherd, M. M. (2006). Research updating the APM Body of Knowledge 4th edition. *International Journal of Project Management, 24*(6), 461.

Müller, R., & Turner, R. (2007). The Influence of Project Managers on Project Success Criteria and Project Success by Type of Project. *European Management Journal, 25*(4), 298.

Nogeste, K., & Walker, D. H. T. (2005). Project outcomes and outputs: making the intangible tangible. *Measuring Business Excellence, 9*(4), 55–68.

OGC - UK Office of Government Commerce. (2007). *Managing Successful Programmes*. Norwich, UK: The Stationery Office.

Packendorff, J. (1995). Inquiring into the temporary organization: new directions for project management research. *Scandanavian Journal of Management, 11*(4), 319–333.

Parikh, M. A., & Joshi, K. (2005). Purchasing process transformation: restructuring for small purchases. *International Journal of Operations & Production Management, 25*(11), 1042–1061.

Pelled, L. H., Eisenhardt, K. M., & Xin, K. R. (1999). Exploring the black box: An analysis of group diversity, conflict, and performance. *Administrative Science Quarterly, 44*, 1–28.

Pennypacker, J. S., & Grant, K. P. (2003). Project management maturity: an industry benchmark. Project Management Journal, March: 4-9.

Pinto, J. K., & Slevin, D. P. (1988). Critical Success Factors across the Project Life Cycle. *Project Management Journal, 19*(3), 67–75.

Pinto, J. K., & Slevin, D. P. (1989). Critical Success Factors In R&D Projects Research Technology Management, 32, 1: 31-35.

PMI Standards Committee (2008). A Guide to the Project Management Body of Knowledge. 4th Edition. Newtown Square.

Rad, P. F., & Levin, G. (2002). *The advanced project management office: a comprehensive look at function and implementation*. Boca Raton, FL: St. Lucie Press.

Rodgers, R., & Hunter, J.E. (1991). Impact of management by objectives on organizational productivity. Journal of Applied Psychology: 322-336.

Rubin, I. M., & Seeling, W. (1967). Experience as a factor in the Selection and performance of Project Managers. *IEEE Transactions Engineering, 14*(3), 131–134.

Scott-Young, C., & Samson, D. (2008). Project success and project team management: Evidence from capital projects in the process industries. *Journal of Operations Management, 26*(6), 749–766.

Shenhar, A. J. (2001). One size does not fit all projects: Exploring classical contingency domains. *Management Science, 47*(3), 394–414.

Shenhar, A. J., & Dvir, D. (2007). Reinventing Project Management: The Diamond Approach to Successful Growth and Innovation. Harvard Business School Press.

Shenhar, A. J., & Dvir, D. (1996). Toward a typological theory of project management. *Research Policy, 25*(4), 607–632.

Shenhar, A. J., Tishler, A., Dvir, D., Lipovetsky, S., & Lechler, T. (2002). Refining the search for project success factors: a multivariate, typological approach. *R&D Management, 32*(2), 111–126.

Simon, Herbert. (1976). *Administrative Behavior* (3rd ed.). New York: The Free Press.

Simon, L. (1995). Understanding the nature of conflicts of interest. http://svc203.wic019v. server-web.com/about-ethics/ethics-centre-articles/ethics-subjects/conflict-of-interest/article-0223.html (Accessed on 16-Feb-10).

Smyrk, J. (1995). *The ITO model: a framework for developing and classifying performance indicators.* Sydney, Australia: The International Conference of the Australasian Evaluation Society.

Snead, K. C., & Harrell, A. M. (1994). An Application of Expectancy Theory to Explain a Manager's Intention to Use a Decision Support System. *Decision Sciences, 25*(4), 499–513.

Tatikonda, M. V., & Montoya-Weiss, M. M. (2001). Integrating Operations and Marketing Perspectives of Product Innovation: The Influence of Organizational Process Factors and Capabilities on Development Performance. Management Science, January, 47.

Tichy, L., & Bascom, T. (2008). The Business End of IT Project Failure. *Mortgage Banking, 68*(6), 28–35.

Turner, J. R. (2009). The Handbook of Project-based Management. 3rd Edition. McGraw-Hill, London.

Turner, J. R., & Muller, R. (2004). Communication and Co-operation on Projects between the Project Owner As Principal and the Project Manager as Agent. *European Management Journal, 22*(3), 327–336.

Veliyath, R., Hermanson, H. M., & Hermanson, D. R. (1997). Organizational Control Systems: Matching Controls with Organizational Levels. Review of Business, winter 1997.

Vroom, V. H. (1964). *Work and motivation.* New York: John Wiley & Sons.

Walker, D. H. T., & Nogest, K. (2008). Performance measures and project procurement. In D. H. T. Walker & S. Rowlinson (Eds.), *Procurement Systems—A Cross Industry Project Management Perspective.* London: Taylor & Francis.

Ward, L. (2000). Project management terms—a working glossary. 2nd ed. ESI International.

Watson, W. E., & Behnke, R. R. (1991). Application of Expectancy Theory and User Observations in Identifying Factors which Affect Human Performances on Computer Projects. *Journal of Educational Computing Research, 7*(3), 363–376.

Wright, J. K., Carpenter, E. W., Hunt, A. G., & Downhill, B. (1958). Observations on the Explosion at Ripple Rock. *Nature, 182*(4649), 1597–1598.

Yiming, C., & Hau, L. (2000). Toward an Understanding of the Behavioral Intention to Use a Groupware Application. Proceedings of the 2000 Information Resource Management Association International Conference, Hershey, PA, USA, Idea Group Publishing: 419-422.

Zwikael, O. (2008a). Top management involvement in project management—a cross country study of the software industry. *International Journal of Managing Projects in Business, 1*(4), 498–511.

Zwikael, O. (2008b). Top management involvement in project management—exclusive support practices for different project scenarios. *International Journal of Managing Projects in Business, 1*(3), 387–403.

Zwikael, O., & Globerson, S. (2004). Evaluating the Extent of use of Project Planning: A Model and Field Results. *International Journal of Production Research, 42*(8), 1545–1556.

Zwikael, O., & Globerson, S. (2006). From critical success factors to critical success processes. *International Journal of Production Research, 44*(17), 3433–3449.

Zwikael, O., Globerson, S., & Raz, T. (2000). Evaluation of models for forecasting the final cost of a project. *Project Management Journal, 31*(1), 53–57.

Zwikael, O., & Sadeh, A. (2007). Planning effort as an effective risk management tool. *Journal of Operations Management, 25*(4), 755–767.

Zwikael, O., Shimizu, K., & Globerson, S. (2005). Cultural Differences in Project Management Processes: A Field Study. *International Journal of Project Management, 23*(6), 454–462.

Index

A

Above-the-line 59, 60, 88, 93–95, 207, 268, 272
Activity 94, 173, 190, 195, 197, 203–204
Appraisal 39, 40, 166, 175–177, 179, 273
Artefact 13, 15, 17, 18, 273
Assessment 38–42, 47, 60, 61, 66, 222, 227, 274

B

Baseline 46, 79, 94, 149, 150, 184, 227, 228, 276
Below-the-line 59, 60, 93, 94, 204, 206, 268, 276
Beneficiary 22, 24, 31–33, 91, 119–121, 276
Benefit 2, 3, 18, 53–55, 57, 58, 60, 61, 175, 268, 276
Budget 59, 165, 173, 201, 204, 205, 217, 254, 255, 276
Business case 62, 136, 138–143, 162–174, 277
  Realised, 42, 69, 267–269
  Approved, 2, 42, 63, 64, 70, 137, 175–177, 246
  Modified, 184, 186, 187, 201, 216
  Original, 3, 200

C

Closeout 89, 244, 245, 260–263, 280
Communication 3, 79, 80, 122–125, 185, 188, 223, 233–236, 281
Complexity 6, 14, 81, 195, 216
Constraint 143–145, 164, 168, 186, 250, 281
Cost 43, 47–49, 58–61, 64, 201–207, 217, 268, 282

Counsellor 100, 102, 107, 108, 120, 220, 221, 256, 270, 283
Customer 12, 22, 26, 28, 31, 32, 119, 120, 151, 153, 162, 264, 283

D

Deadline 21, 182, 191, 192, 196
Deliverable 3, 13, 93, 284
Disbenefit 18, 53, 54, 58–61, 175, 176, 267, 268, 285

E

Earned value 240, 286
Evaluation 39–41, 45, 262–268, 286

F

Fitness-for-purpose 91, 147, 161, 288
Funder 2, 22, 28–30, 58, 119, 267, 289

G

Gantt chart 196, 198, 203, 204, 289
Governance 9, 28, 91, 92, 95–102, 109, 140, 165, 171, 172, 185, 290

I

Impactee 119, 120, 165, 171, 187, 292
Input-Process-Output (model) 15, 17, 22, 23, 293
Input-Transform-Outcome (model) 22–30, 61, 74, 91, 120, 294
Investment 2, 39, 40, 65, 135, 175
Issue 237, 294

I (*cont.*)
  Management, 186, 237, 238, 246
  Life-cycle, 237–239
  Register, 41, 174, 186, 237, 238
  Report, 165, 174, 253

L
Likelihood 116, 208, 211–213, 296
Leader 109, 117, 210, 238, 240–243

M
Measurability 17, 149, 297
Memorandum of Understanding 29, 99, 106,
    231, 297
Methodology 6, 30, 297
Milestone 87, 173, 185, 191, 198, 200, 203,
    251, 297
Mitigation 92, 140, 156, 174, 179, 208–210,
    213, 214, 248, 312

O
Objective statement 147, 149, 160, 299
Operational 13, 14, 16, 21, 26, 299
Operations 1, 13, 26, 89, 90
Outcome 17, 18, 148–151, 261, 265, 299
  Desirable, 11, 17, 29, 286
  Detrimental, 48, 286
  Fortuitous, 17, 34, 63, 66, 262, 289
  Natural, 23, 24, 260
  Realisation, 85, 88, 89, 259, 260, 299
  Synthetic, 23, 260
  Target, 2, 17, 19–23, 56, 57, 64, 90,
      148–150, 152, 319
  Undesirable, 41, 43, 46, 48, 49, 54, 58, 64,
      165, 171, 187, 322
Outlay 43, 60, 199, 239, 246, 299
Output 17–19, 152, 161, 162, 244, 299

P
Performing organisation 79, 97, 189, 301
Planning 14, 85, 87, 88, 90, 181, 302
Plausibility 149, 247
Portfolio 4, 6, 92, 132, 133
Principle 96, 102, 104, 106, 202
Process 1, 13–15, 21–23, 76, 303
Programme 6, 125–132, 164, 304
Project
  Administrator, 33, 101, 233, 305
  Adviser, 100, 102, 106, 107, 120, 220,
      256, 269, 273

Appraisal, 39, 40, 166, 175–177, 179, 273
Assessment, 38–42, 47, 60, 61, 66, 222,
    227, 274
Attractiveness, 53, 67, 92, 113, 133,
    175, 176, 179, 180, 275
Champion, 87, 90, 109, 117, 120,
    135–137, 279
Counsellor, 100, 102, 107, 108, 120,
    220, 221, 256, 270, 283
Cost, 43, 47–49, 58–61, 64, 201–207, 217,
    268, 282
Customer, 12, 22, 26, 28, 31, 32, 119, 120,
    151, 153, 162, 264, 283
Descoping, 192, 248, 249
Environment, 5, 7, 9, 27, 28, 85, 90, 226,
    234, 253, 306
Evaluation, 39–41, 45, 262–268, 286
Infeasible, 8, 144, 191–192, 204, 205,
    249, 292
Manager, 9, 28, 40, 86, 90, 189, 233,
    264, 307
Owner, 28–31, 62, 90, 102–104, 216, 247,
    265, 308
Plan, 3, 184, 185, 219–222, 308
Project Management Office 82, 110, 111, 307

Q
Quality 42, 47, 50, 185, 188, 193,
    221, 222, 309

R
Rank 53, 56, 179, 216–219
Regression test 41, 42, 63–69
Risk 177–180, 207–213, 236, 237, 249, 311
  Exposure, 24, 67, 174, 178, 180, 208,
      212–214, 312
  Manager, 233, 236, 312
  Register, 174, 186, 214, 215, 313
  Riskiness, 53, 63, 67, 92, 178, 313
  Report, 253, 313

S
Satisficing 57, 179
Schedule 91, 196, 203, 239, 250
Scope 23, 90, 91, 112, 145, 146, 247–250, 314
  Scoping statement, 146–148, 153, 160,
      164, 165, 187, 314
Severity 130, 174, 179, 209, 211, 212, 315
Sponsor 12, 110, 315
Stakeholder 3, 8, 91, 110–125
  Classes, 101–108, 119, 120

Engagement plan, 117, 118, 124, 316
Statement of objective 127, 147, 149, 160,
        298, 317
Steering committee 100–102, 220, 251–255,
        269, 319
Success 37, 69–76, 256
    Project investment success, 65–69, 267, 269
    Project ownership success, 62–65, 194, 267
    Project management success, 42–53, 194

T
Task 14, 195–197, 202, 203, 273, 319
Team 3, 102, 105, 106, 214–216,
        245–247, 319
Template 143, 164, 165, 185–187,
        253, 254, 320
Threat 140, 174, 178, 208–213, 320
Timeframe 42, 49, 192, 321

Trade off 31, 40, 43, 46, 250
Trigger 1, 40, 87, 208

U
Utilisation 22–26, 90, 91, 264, 265, 322
    Utilisation map, 132, 146, 153, 158–161,
        170, 266, 322

V
Value 18, 22, 53, 54, 56, 57, 61, 288

W
Work Breakdown Structure 185, 195–198, 322
Worth 53, 54, 59, 60, 63, 64, 268, 309

Printed in the United States
By Bookmasters